ISLAMIC
FUNDAMENTALISM

ISLAMIC FUNDAMENTALISM

An Annotated Select Bibliography of Sources in English

KHALID KAMAL FARUQI

M.A., M. Lib. Sc. (Alig.)
Deptt. of Library & Information Science
Jamia Millia Islamia
(Central University)
New Delhi - 110025

CBS

CBS PUBLISHERS & DISTRIBUTORS

4596/1-A, 11-DARYAGANJ, NEW DELHI-110002

Assisted by
MMA Ansari, Asstt Librarian
Dr. Zakir Husain Library
Jamia Millia Islamia,
New Delhi - 110025.

ISBN : 81-239-0447-9

Published by:
Satish Kumar Jain
CBS Publishers & Distributors
4596/1A, 11 – Daryaganj, New Delhi - 110 002

Typeset at:
Phoenix Computer Centre, New Delhi-110 002.

Printed at:
R.V. Printers, Bhola Nath Nagar, Shahdara, Delhi.

FOREWORD

Fundamentalism & "Fundamentalist" are the words which were orginated in the West against the Revivalists among the Christian World during the medieval period.

There is nothing like "Fundamentalism" or "Fundamentalist" in Islam. There is no criterion to classify Muslims into Fundamentalits or Non-Fundamentalists. Every Muslim if he/she is at all a Muslim has to adhere to the fundamentals of Islam Viz. - al-Kalimat-al Tayyiba, Salat, Zakat, Saum, Haj; and other Islamic principals and teachings. The term "Fundamentalism" or "Fundamentalist", in the context o Islam and Muslim, have been fabricated to defame such Muslims who want to practice Islam, hence such terms are quite alien to Islam. As a matter of fact when the western authors saw the revivlism of Islam in some of the Muslim countries during the last two decades, they were frightened with its popularity, hence they started to shed hatred against Islam and the Muslims by branding them as "Fundamentalists". I would like to quota Sultan Shahin in this connection. He has given a clear picture in his article "Islam And The Jamahiriya: Two facts of the struggle against Facism" (*Secular Democracy*; 14, 5; 1981, May; 36-43): "When the Islamic slogans are (i.e. were) raised by anti-imperialist and anti-zionist movements, as .was the case in Egypt, Iraq, Syria, Libya and later in Iran, the imperialists see (i.e. saw) a mortal danger to western civilization and launch (i.e. lanched) a crusade against Islam. The Iranian people's revolt was more than just an ordinary uprising against tyranny. It signified the refusal of a whole nation to accept the model of social and economic, development imposed on it from abroad."

Mr. Khalid Kamal Faruqi has done a commendable work by collecting and compiling a lengthy bibliography on "Islamic Fundamentalism" and related topics in English language. He has done an objective study and has presented in this book almost all the works-articles and books, whether pro-Fundamentalism or against. Moreover, he has also divided these works under various headings and sub-headings like Islamic Fundamentalism, concepts; Islamic Fundamentalism in relation to Modernism etc. He has also given the position of various countries (like Iran, Egypt, Libya, China, Indonesia etc.) and regions (like Middle East, Africa etc.).

He has also given various aspects of the subject. Further he has also covered works of both Oriental and Western authors. Thus this book is a honest "Research work" done by Mr. Khalid Kamal Faruqi. I hope this book would be quite helpful and beneficial to the researchers in this field.

(Majid Ali Khan)
Prof. and Head Deptt of Islamic Studies

Jamia Millia Islamia,
New Delhi.
24.08.95

INTRODUCTION

Fundamentalism as a term is a contribution of the Christendom to the world of ideas. It originated with the rise of a group of such Christians in the United States, who sought to reject all attempts, made formerly or being made currently, to interpret Christian beliefs to meet the exigencies of the changing world, and insisted on a literalist approach toward the correct understanding of the holy Bible. The group also declared that the truth as contained in the Biblical verses and as understood and explained by the Church, since its very inception, could not be challenged.

It is not the proper place to discuss the various trends of through about the revealed nature, compilation, and the most authentic and genuine text of the Bible itself, which have been subjects of religio-intellecual controversies among scholars for centuries. This confusing situation in the Christian world has led to its moral and spiritual foundation having been shaken to the very core. No doubt, sometime people of religious conscience raise their voice against the unbridled materialism and too much secularization of life and suggest that if the Biblical teachings are followed in good faith as the fundamentals of thought and action, the West could still be morally and spiritually strong along with the superiority in science and technology. But such a voice is immediately condemned as an advocacy of fundamentalism.

Now in recent Year, attempts made by some Islamic movements in the world of Islam to reform their societies in conformity with the basic principles of Islam have been dubbed by the Western media as reactionary and retrogressive as, in their view, these ultimately verge on their definition of *fundamentalism*. Nay, these are propagated and publicised exactly as such i.e. as the rise of 'Fundamentalism'.

Islamic fundamentalism as it is mischievously made to be understood, in fact, owes its origin to the ideas and ambitions which plead for political and economic expansionism of the West in the world of Islam. It has nothing to do with the Islamic awakening which represented a natural and normal historical process, first, of protest against and liberation from the political domination of the West and, then, of a desire to rehabilitate Islam as a life-force in the areas of its influence.

The West was, however, not prepared to give the Muslim countries a reprieve in order to organise their affairs themselves after the end of the colonial era. Its plenty and prosperity depended on the continuation of its dominant role, in the form of its military presence and economic influence, in Muslim countries. Had it allowed them freedom to rebuild themselves and be the masters of their own destiny by purposefully utilizing the deversified assistance from the various agencies of the United Nations, there, would, perhaps, not been so much resentment on their part against the West, western civilization and all that it stands for, good or bad.

But, the West wanted exactly oppsite to what the Muslims aspired for. Hence, overtly or covertly, a confrontation, in which the West has, so far, succeeded in contriving to have the full support of the local establishment, generally in control of the Westernised or pro-West elements with their own vested interests threatened, in turn, by the gust of Islamic resurgence.

Islamic fundamentalism has, thus, an entirely political connotation. It cannot be explained or made to be understood in terms of 'orthodoxy' which represented the middle and straigth path in classical Islam. Based on the *Qur'an* and the *Sunnah* and unhampered by the shackles of *taqlid* as it later developed, it was more open and generous in assimilating foreign elements and learning and accepting gainful sciences from whoever possessed them, and more flexible in interpreting the scriptures to meet the axigencies of time and space. Return to classical Islam is being liberal rather than being fundamentalist. But, the West, because of its subjection to its religio-intellectual history of the past centuries, is unable to see and appreciate Islam as a liberating force and a champion of peace and progress.

The West claims to stand for progress and democracy. It has become more clamarous for these ideals after the end of the cold war. It raises the bogey of Islamic fundamentalism whenever and wherever it suits it to protect its vested interests in the name of progress and enlightenment; it, however, leaves no chance to support the undemocratic leaders in Muslim countries, who are sometime found to sympathise with fundamentalism in order to legitimize their own authority; sometime it is also used by the vested-interests as an ideological tool against the liberal trends in Islam. On all such fronts the West is their to serve its clients i.e. the exploiting ruling class.

Mr. Khalid Kamal Faruqi has done a good job in bringing together a substantial portion of the vast material available on the various aspects of Islamic fundamentalism. Though selective, the study well represents different periods and schools of thought, and provides a useful bibliography, with intelligent and readable annotations, on the subjects. It certainly provides a base for further research and discussion in the field.

It is hoped that this would be received well, read and discussed by the students and scholars both.

(ZIYA-UL HASAN FARUQI)
Prof. & Ex-Director,
Dr. Zakir Husain Institute of Islamic Studies
Jamia Millia Islamia
New Delhi-110 025.

Zakir Nagar,
New Delhi
5.8.95

PREFACE

Perhaps one of the main events in the twentieth Century which has attracted such a world wide attention is the Islamic Resurgence popularly known as Islamic fundamentalism. It menace the existing political order and stability in a number of Muslim Countries. The scholar world over, particularly from those in the west have to justify this phenomenon in terms of their favourite theories. Despite the powerful impact and growing interest in "Islamic fundamentalism", there is no proper guide to the rese-arch material available.

AIM AND SCOPE

The present study is intended to bring all significant literature that is available on the "Islamic Fundamentalism" in the form of annotations at one place. Although the present bibliography is selective in nature, an attempt has been made to cover all the aspects of "Islamic Fundamentalism" and to make it fully representative of the various periods and school of thought.

Much has been written about the subject and still more will be written about it. This is a very small attempt to bring together some of the vast material on the subject which will provide a base for further research, discussion and comprehensive analysis of the subject.

METHODOLOGY

While starting with this task a general survey of literature available in important libraries, viz Dr. Zakir Hussain Library, Jamia Millia Islamia, New Delhi, Indian Council of World Affairs Library New Delhi, India International Centre Library, New Delhi, Delhi University Library and Maulana Azad Library at Aligarh was made. Out of the total number of periodicals covering the field, only important one, but of different school of thought, was selected for this bibliography. A list of periodicals and books documented has been given.

ARRANGEMENT

SUBJECT HEADING

Efforts have been made to arrange the entries under the co-extensive subject headings. For this purpose a comprehensive list of subject headings was compiled. Although there is always scope for difference of opinion on any issue, the list of subject headings will generally be found following a coherent sequence.

Under the specific subject headings the entries have been arranged alphabetically by author. The entries are serially numbered.

ANNOTATION

For the reader who does not have the opportunity to consult all the sources available in this bibliography, the annotations should provide adequate information on the sources. An attempt has been made to bring out in each annotation the essential points of the article. In many cases the author's point of view has been given in his own words.

SELECTION OF ARTICLE—CRITERION

The selection of articles included was made on three criteria. (*a*) that an article should contribute to general knowledge about the "Islamic fundamentalism (*b*) the article should throw light on some issue or problem related to Islamic fundamentalism and (*c*) the article should project Islamic fundamentalism from different angle i.e., Islamic, Marxist, Modernization. The bibliography has focused on material published only in English language due to limitation of space.

INDEX

The bibliography contains Author and title index in alphabetical sequence. Each index guides to the specific entry or entries in the bibliography.'

I hope the users will find this bibliography very useful research tool and will contribute towards a better appreciation understanding of Islam. The conclusion to be drawn from this bibliography are left to the readers.

— Khalid Kamal Faruqi

ACKNOWLEDGEMENTS

Contributions from many institutions and people have made this bibliography possible. I must thank all those people who directly or indirectly contributed to the preparation of this book.

Grateful thanks are due to Prof. S. Ansari, Ex-University Librarian of Jamia Millia Islamia and Director, Deptt. of Library and Information Science, Jamia Millia Islamia (Central University), New Delhi - 110 025 for his help incouragement and patronage throughout my carrier.

I am also thankful to my respected teachers Prof, Zia-ul-Hasan Faruqi for his encourgement and affection by writing - Preface - for my book.

I must express my thanks to Prof. Majid Ali Khan, Head, Deptt. of Islamic Studies, Jamia Millia Islamia, New Delhi who gladly agreed to write foreword for my book.

I am also grateful to Prof Agha Ashraf Ali, Ex-commissioner of Education, J&K States, Srinagar for giving constructive suggestions for improvement and correction of the manuscript.

Grateful special thanks are also due to Prof. Mohd. Azhar Ansari, Ex-Professor & Head, Deptt. of History, Jamia Millia Islamia for his incouragement in my carrier and indirecting contributing to the Publication of this book.

Grateful thanks are also due to my colleague Mr. M.M.A. Ansari, Mr. Rabiullah Khan, Mr. Gufran Quaraishi, Mr. Mohd. Ahmad and Mr. G.D. Sharma for their help in various form in brining out this book.

Grateful thanks are also due to Mr. Jain of CBS Publisher and their representative Mr. B.R. Sharma who have taken great pain in brining out this book.

In the end I want especially to thank my wife TALAT and my daughters KAKUL and AMBER whose love, inspiration encouragement and cooperation made it possible to complete this study.

Finally I dedicate this book to ABBU and AMMI, my parents.

— Khalid Kamal Faruqi

List of journals and books documented

A. Journals

1. African Perspectives (Netherlands)
2. Arabia (Saudi Arabia)
3. Asian Affairs (London)
4. Asian Studies (Philippines)
5. Asian Survey (California)
6. Australian Outlook (London)
7. British Society for Middle Eastern Studies Bulletin (Oxford)
8. Bulletin of Christian Institute of Islamic Studies (Hyderabad)
9. Bulletin of the School of Oriental and African Studies (London)
10. Current Affairs Bulletin (Sydney)
11. Current History (Philadelphia)
12. Economic and Political Weekly (Bombay)
13. Far Eastern Economic Review (Hong Kong)
14. Hamdard Islamicus (Karachi)
15. IDSAJ (Institute for Defence Studies and Analysis, New Delhi)
16. India International Centre Quarterly (New Delhi)
17. Indian Express (New Delhi)
18. International Journal (Toronto)
19. International Journal of Middle East Studies (Cambridge)
20. International Studies (New Delhi)
21. Iranian Studies (Boston)
22. Islam and the Modern Age (New Delhi)
23. Islamic Culture (Hyderbad)
24. Islamic Order (Karachi)
25. Islamic Quarterly (London)
26. Islamic Studies (Islamabad)
27. Jerusalem Quarterly (Jerusalam)
28. Journal of Anthropological Research (Albuquerque)
29. Journal of Commonwealth and Comparative Politics (London)
30. Journal of Dharma (Bangalore)
31. Journal of South Asian and Middle Eastern Studies (Villanova)

32. Journal of the Economic and Social History of the Orient (Leiden)
33. Link (New Delhi)
34. Mainstream (New Delhi)
35. Man (London)
36. Man and Development (Chandigarh)
37. Manthan (New Delhi)
38. Middle East Journal (Washington)
39. Middle Eastern Studies (London)
40. Millennium (London)
41. Month (London)
42. al-Mushir (Rawalpindi)
43. Muslim Digest (Durban)
44. Muslim World (Karachi)
45. Muslim World League Journal (Mecca)
46. New East (Jerusalam)
47. New Quest (Pune)
48. Orbis (Pennsylvania)
49. Pecific Affaris (Canada)
50. Radiance (New Delhi)
51. Readers Digest (Bombay)
52. Religious Studies (New York)
53. Review of Religious Research (Kansas)
54. Secular Democracy (New Delhi)
55. Seminar (New Delhi)
56. Social Campass (Belgium)
57. Social Scientist (Kerala)
58. Stretegic Analysis (New Delhi)
59. Studia Islamica (Paris)
60. Studies in Islam (New Delhi)
61. Theology Today (Princeton)
62. Third World Quarterly (London)
63. Times of India (New Delhi)
64. Voice of Islam (Karachi)
65. Washington Quarterly (Washington)

B. Books

1. ABDUL HAMEED, *ED*. Islam at a glance. New Delhi, Indian Institute of Islamic Studies, 1981.

2. ANWAR MOAZZAM, *ED*. Islam and contemporary Muslim World, New Delhi, Light and Life, 1981.

3. ESPOSITO (John L), *ED*. Islam and development. New York, Syracuse University Press, 1980.

4. ESPOSITO (John L), *ED*. Voices of resurgent Islam. New York, Orford University Press, 1983.

5. KHURSHID AHMAD, *ED*. Studies in Islamic economics. Leicester, Islamic Foundation, 1980.

6. KHURSHID AHMAD AND ANSARI (Zafar Ishaq), *ED*. Islamic Perspectives. Leicester, Islamic Foundation, 1979.

7. LEWIS (I M), *ED*. Islam in Tropical Africa. London, International African Institute, 1980.

8. MOHAMMED AYOOB, *ED*. Politics of Islamic Reassertion. New Delhi, Vikas, 1982.

9. MUHAMMAD HAMID AL-AFENDI and NABI AHMAD BALACH, *ED*. Curriculum and Teacher education, Jeddah, King Abdul Aziz University, 1980.

10. PISCATORI (James P), *ED*. Islam in the political process. Cambridge, Cambridge University Press, 1983.

11. SALEM AZZAM, *ED*. Islam and contemporary society. London, Longman, 1982.

12. SIDDIQUI (Kalim) and othes, *ED*. Islamic Revolution. London, Open Press, 1980.

13. STODDARD (Philip H) and others, *ED*. Change and the Muslim World. New York, Syracuse University Press, 1981.

14. UTAS (Bo), *ED*. Women in Islamic societies, London, Curzon Press, 1983.

B. BOOKS

1.

2.

3.

4.

5.

6.

7.

8.

9.

10.

11.

12.

13.

14.

CONTENTS

BIBLIOGRAPHY

1. ABDUL MOGHNI. Islamic fundamentalism: What it means. *Muslim Wld League J*; 10, 10; 1983, July-Aug; 15-17.

   Islamic fundamentalism appears to have become, of late, the whipping boy of the modernists. Whenever whatever they want to condemn in Islam, as a religion, they at once dub it as fundamentalism, which they have raised as bogey. But the fact is that fundamentalism in Islam is nothing else than what is known as orthodoxy. This is some thing different from simple conservatism. It is also not fanaticism, in the western sense. Even puritanism of the Christian connotation cannot define Islamic fundamentalism. The term orthodoxy itself is not the same in the history of Islam as in that of Christianity.

2. AGWANI (M S). Islamic fundamentalism: Myth and reality. *Man Develop*; 2, 3; 1980, Sept; 85-91.

   Author says that "the phenomenon of Islamic resurgence must be seen in its proper perspective". Since Shah left for the exile, the western Media have scare stories about Islam. In western writings, the hypothesis on the so called fundamentalist reaction rest on two premises. First, the Muslims are being swept by anti-modern forces of religious revival. Second, its success might affect the supply of oil. Suggests that "in assessing the present day potentialities and limitation of Islamic fundamentalism it is necessary to trace its origin and impact in early history of the Islamic body politic".

3. BARAKAT AHMAD. Myth of resurgence. *Seminar*; 290; 1983, Oct; 14-18.

   In this article, the author raised a question "What is Islam?" He wrote that the word 'Islam' is used with at least three different meaning and discussed it in length. He also says that "terms such as "orthodoxy", "fundamentalism", "traditionalism" and "rationalism" are alien to Islam." But "all Muslims are fundamentalist". It is therefore nonsense to classify Muslims as fundamentalists and non-fundamentalists. There was no period of decline when a mujaddid (renewer) was not providentially sent to arrest decay and pull back the community to a closer contact with the principles of the Quran and the Sunna. He has given so many names of the Mujaddid of Islam from different period. But the "present upsurge is not religious, it is political though formulated in a religious language." Hasan al Bann's Ikhwan al Muslimun or Maulana Maududi's Jama'at-i-Islami, the pro-Khomeini Fida'iyan-i-Islam, the organization of Algerian Ulama, the National Salvation Party in Turkey are working to capture political power.

4. MITRI (Tarik) and SCHOEN (Ulrich). Fundamentals and fundamentalism. *al-Mushir*; 6, 3; 1980; 137-149.

The author briefly characterize fundamental aspects of Islam as *iman* (faith, *tawakkul* (trust), *din* (the totality of rules of life) and *umma* (community of Muslims) gives further a historico-philosophical typology: theocratic, apocalyptic, "historical" (for instance a Marxist who 1. takes seriously the autonomy of the laws of nature and history and 2. does not want to betray Islam) and mystical types of Islam. As for "Islam and modernism": the Quran fully acknowledges scholarship and science; the real problem for Islam starts with social sciences. *Final thesis*: three theoretical models for thinking and action in the Islamic world are a. a fundamentalistic and retrospective view; b. a complete openness to modernity, western modernity serving as standard of the specific and universalistic elements of Islam. The answer to the question: "What kind of Islam, what kind of modernity? Still needs to be found.

5. WATT (W Montgomery). Contemporary political relevance of the religion of Islam. *Scottish Relig Stud*; 1, 2; 1980; 85-96.

The author points out that the recent Iranian revolution has highlighted the Islamic resurgence, which he attributes to a growing sense of insecurity, itself caused by the spread of western technology and ideas in Islamic countries. The main political trust of the Islamic resurgence is negative--an attempt to get rid of the west's impact while using its technology, and the religious institution in the Islamic world is both influential and conservative. In a brief historical survey the author outlines the role of Sharia religious-legal code interpreted by the Ulema, a religious body whose influence dwindled in the Ottoman Empire after about 1850, the Sharia Courts being replaced by Ataturk's European style code in 1920's. The Ulema having involuntarily relinquished power "the combination of a resurgent mass-religion and a religious institution seeking to recover power has produced the dynamic behind the Islamic Revolution of 1979 in Iran". Economic discontent also was focussed, mainly negatively, against the un-Islamic U.S.A. Iran however is unique among the Muslim countries in that the official religion is Shi'ism according to those doctrine, the 'hidden' Imam is believed superior to any temporal government. The other main Islamic countries are Sunnite, and the religious institution has traditionally been subordinate to the ruler. The author concludes with assumptions of Islam in the areas of education; the adaptation of the Sharia to the modern world; the relations between Muslims and non-Muslims. He identifies an urgent need for Muslim scholars to clarify the position of the Sharia on many pressing contemporary issues. "What I would hope for in the future is that those I have called liberals might have some success in directing the energies of the resurgence into more positive channels".

ISLAMIC FUNDAMENTALISM, Afghanistan

6. ANWAR MOAZZAM and MOHAMMAD AHMEDUL-LAH. Afghanistan: Revolution and revivalism. *In* Anwar Moazzam, *Ed. Islam and contemporary Muslim world.* New Delhi, Light and Life, 1981, p. 85-106.

One is forced to study the recent developments in Afghanistan as a struggle between two socio-political ideologies and systems rather than a mere internal problem, thanks to the wave of "Islamic resurgence" sweeping through West-Asia, particularly the neighboring Iran and Pakistan. The author has traced the development which are basically political in this article.

Concludes: "In fact, they need not do so since the socio-economic goals of socialism and Islam are, to a large extent, co-terminous. It is a socio-economic egalitarianism of Islamic tenets, projected as 'Islamic socialism' by Nasser of Egypt, Bhutto of Pakistan and the Bath Party of Syria and Iraq, which is the greatest safeguards against the acceptance of communism in Muslim lands."

ISLAMIC FUNDAMENTALISM, Africa

7. KASULE (Omar Hasan). Islam' reawakening in Africa: Prospects and challenges. *Radiance*; 6, 49; 1981, April, 193; 3.

   The Islamic Movement started 1400 years ago with the migration of Prophet Mohammed expanded beyond the Arabian peninsula. Later decline set in and Muslims who at one time led the world in knowledge and science, stagnated, were overtaken by the western world. To regain their former status Muslims all over the world are calling for the return of Islam. This article seeks to examine this Islamic reawakening from an African context and to assess its future prospects. The central theme of the article is that Africa is a ready and fertile ground for Islam and the Islamic workers should give it the attention it deserves.

ISLAMIC FUNDAMENTALISM, Africa and Middle East

8. DEKMEJIAN (Richard Hrair). Islamic revival in the Middle East and North Africa. *Current Hist*; 78, 456; 1980, April; 169-174, 179.

   After discussing the phenomenon of Islamic revival in details, the author comes to the conclusion "that Islamic revivalism has become an important factor in the politics of the Middle East, North Africa and South-west Asia. It is, however, still in its nascent stage". Author declares that despite similarities, the concrete manifestation and consequences of Islamic revival are likely to be different in specific Islamic countries. Author also feels that the "anatomy of Islamic fundamentalism may be shaped by unforeseen events that will surely buffet the Islamic conscience in the coming decade".

ISLAMIC FUNDAMENTALISM, Algeria, History

9. VATIN (Jean-Claude). Popular puritanism versus state reformism: Islam in Algeria. *In* James P. Piscatori, *Ed. Islam in the political process*, Cambridge, Cambridge University Press, 1983, p. 98-121.

   "There have been many signs of an Islamic revival in Algeria during the last ten years. This is largely attributed to two different factors. The first is the comprehensive character of Islamic fundamentalism". The second factor is more localized. This revival has been characterised more by an increase in religious fervour, of religiosity than by the development of religious belief or spirituality. People, in Algeria, are turning or returning to Islam because they are looking for answers to the problems of today. Although there was no significant political instability, people were anxious about the future. In such a situation "a puritanical, revivified Islam could be effectively used for criticizing official actions, while providing a political lever for popular and not just traditionalist movements". The orthodox or conservative fundamentalist have existed in Algeria for some time. But the state has other means at its disposal for countering the problems of Islamic criticism.

ISLAMIC FUNDAMENTALISM, America

10. HADDAD (Yvonne Y). Islam in America: A growing religious movement. *Muslim Wld League J*; 9, 9; 1982, July; 30-34.

It is obvious to observers of the American scene that Muslims are playing an increasingly important role in America. Their ranks include eminent doctors, engineers, professors, and scientists who participate in the shaping of the future of America. Islam has appealed to a large spectrum of Afro-American, including athletes and entertainers. Muslims opinion and views are voiced in various publications. The federal government has taken note of the conversion of inmates in prisons by providing the new-Muslims with special places for the practical of the Jumma prayer. Halal food that does not contain ham or pork by-products is prepared for them and they are allowed to fast during the month of Ramadhan, with meals served at appropriate times. The American Navy has sought assistance from Muslim leaders in the selection of books on Islam to be distributed to chaplains to help them relate to the growing number of Muslim men enlisted in the armed forces.

ISLAMIC FUNDAMENTALISM, Causes

11. BENNABI (Malek). Islam in history and society. *Islamic Stud*; 18, 1; 1979; 33-47.

The author criticizes, among other, HAR Gibb's thesis that "atomism—this turn of mind incapable of generalizations—is a special characteristic of the Arab mind". It is rather a question of the modality of human spirit in general when it has not yet attained a certain degree of development and intellectual maturity, or has passed it by. More precisely, the discursive spirit inscribes itself, in the historical evolution, between two stages of atomism" (ep. Europe in the pre-Cartesion period or the post Khaludnian period in the Muslim world). The author further deals with: a. "the cyclical phenomenon": he condemns the habit of regarding in isolation a phenomenon "civilization" and a phenomenon "decadence". He finds that on this point the Muslim world is in need of clear ideas that would guide its present effort of renaissance. The era of decadance which commenced with the post al-Muwahhid man (al-Muwahhidun ruled over North Africa and Spain from 1130-1269 A.D.). The dynamism of the Arab world explains the extreme rapidity of the expansion of Islam, giving to a world dominated by individualism a sense of collective that determined its historical orientation.

12. DEKMEJIAN (Richard Hrair). Anatomy of Islamic revival: Legitimacy crisis, ethnic conflict and the search for Islamic alternative. *Mid East J*; 34, 1; 1980; 1-2.

The recent regeneration of the Islamic ethos appears to have caught the non-Islamic world by surprise. To a western world preoccupied with growing economic problems and security concerns. The new challenge of Islam appears disconcerting and even ominous. Few were those, both in the west and in the communist world, who were able to anticipate an Islamic resurgence in the modern context. This essay is a preliminary attempt to identify some of the characteristics of the contemporary Islamic revival and to discern the possible causal factors responsible for its emergence.

13. ENGINEER (Asghar Ali). Islamic fundamentalism. *New Quest*; 23; 1980, Sept-Oct; 295-300.

Describes the "recent upsurge in the Muslim world" (that) has thrown the question of Islamic fundamentalism into bold relief. The western media have shown great interest in the events. However one is greatly disappointed if one expects unbiased and objective reporting by these agencies. Before discussing the phenomenon of Islamic fundamentalism, the author has discussed the role played by the media. He also says that these developments of religious revivalism cannot be divorced from political motives. Because in none of these countries has 'Islamisation' of the society been the result of a people's movement. Thus crux of the matter is political power and not concern for religion.

14. ESPOSITO (John L). Islam and Muslim politics. *In* John L Esposito, *Ed. Voices of resurgent Islam*, New York, Oxford University Press, 1983, p. 3-16.

Purpose of this paper is to provide some insight into a phenomenon that has swept the Muslim world. It has been described by various titles: Islamic Resurgence, Islamic Revival, Militant Islam, Rise of Islamic fundamentalism. This paper will explore the origins, development, prospects, and implications of the Islamic resurgence through both analytical studies (on the history and major influences of the resurgence) as well as the statements of Muslim Scholars/activists engaged in the political social and economic aspects of the Islamic resurgence today.

15. FAZLUR RAHMAN. Roots of Islamic neo-fundamentalism. *In* Philip H Stoddard and others, *Ed. Change and the Muslim World*, New York, Syracuse University Press, 1981, p. 23-35.

Author points out that "the manifestation of contemporary Islam commonly referred to as Islamic fundamentalism is a product of the interaction of long historical development in Islam.

Fundamentalism must be seen in this double perspective and not interpreted merely as an Islamic reaction to the West in the light of newly gained political independence". Author says "within the Islamic world there are certain common responses to the phenomenon of Islamic revivalism". The author discusses fundamentalism and its various facets in details. He also compared and contrasted between traditionalists and neo-fundamentalists. He also discussed the shortcomings of neo-fundamentalism and Islamic future.

16. KHURSHID AHMAD. Nature of the Islamic Resurgence. *In* John L Esposito, *Ed. Voices of resurgent Islam*, New York,

Oxford University Press, 1983, p. 218-29.

Opine that "the Islamic resurgence is primarily an internal, indigenous, positive and ideological movement within Muslim society. It is bound to come into contact, even clash with forces in the international arena. The close contact of the west, particularly through colonial rule is relevant but not the most decisive factor in producing the Islamic response." "There is nothing pathologically anti-western in the Muslim resurgence". Its goal is a "restructuring of society, the re-building of socio-economic life on the foundations of Islam." "Amidst the conflict of western and Islamic culture, the resurgence represents a third alternative to capitalist and socialist system, a positive creative response to the challenge of modernity".

17. SULTAN SHAHIN. Islam and the Jamahiriya: Two facts of the struggle against facism. *Secular Democracy*; 14, 5; 1981, May; 36-43.

Author is of the opinion that "there are conflicting reasons for the present-day regeneration of Islam". On the one hand, it finds expression in the Muslim people's anti-imperialist solidarity. On the other hand, attempts are being made to use that movement as an instrument of neo-colonialism and religious bounds, and contrapose it to the progressive and democratic forces. Islam has always been exploited by the rulers for their own selfish ends. "As the class struggle sharpened Islamic slogans were from time to time advanced to impart a religious character to social and national liberation movements. Imperialism tries to show that the world's main contradiction—between socialism and capitalism—is a struggle between believers and non-believers. When Islamic slogans are raised by anti-imperialist and anti-zionist movements, as was the case in Egypt, Iraq, Syria, Libya and later in Iran, the imperialists see a mortal danger to western civilization and launch a crusade against Islam. The Iranian people's revolt was more than just an ordinary uprising against tyranny. It signified "the refusal of a whole nation to accept the model of social and economic development imposed on it from abroad". "There would appear to be no contradiction between Islam and the third universal Islamic theory as explained in the Green Book by Col. Qadhafi". Of greater contemporary, relevance, however, is the fact that the Third Universal Islamic Theory has now been fully implemented in the

Jamahiriya and as Libya was only the first field of experiment for the theory, attempts to implement it can now be made elsewhere. Hence the immense popular interest in the Theory and its actual practice in the Libyan Arab Jamahiriya. The author also discusses the Islamic system of Libyan Arab Jamahiriya and its economic implications. The author also gives references of great reformers. "The late 19th and early 20th century, for instance, witnessed the emergence of Jamaluddin Afghani and Sheikh Mohd Abduh who left an indelible impression on the Muslim mind throughout the world." The author is of the opinion that "Col Quadhafi in his Third Universal Islamic theory has given a new interpretation of Islam with a view to ameliorating the plight of the common man.

ISLAMIC FUNDAMENTALISM, Causes, Asia

18. DIL (Shaheen F). Myth of Islamic resurgence in South Asia. *Current Hist*; 78, 456; 1980, April; 185-86.

"In attempting to discover an indigenous and effective social structure, Third World nations are finding that all that is left to them their religion and their remote past". What is happening in Iran, Pakistan, Bangladesh, Afghanistan, Kabul and so many other Muslim countries "is not a resurgence of Islamic fervour leading to jihad, or holy war against the infidels, but rather a desperate attempt at self-definition in a world where the powerless are too often defined in terms of the powerful". The author warns that "it would be grave error for American foreign policy makers to misinterpret this phenomenon. For too long the West did not pay sufficient attention to the political power of Islam. Then, with the oil crisis and the revolution of Iran, political analysts suddenly reversed course and began to talk of the resurgence of a hostile and fundamentalist Islam on a global level"

ISLAMIC FUNDAMENTALISM, Causes, Egypt

19. RIZVI (S Ameenul Hasan). Back to religion in Egypt. *Radiance*; 9, 21; 1971, Dec, 5; 2.

Author says that with the break up Ottoman Empire, almost all the Muslim countries of West Asia and Africa were swayed by a tidal wave of territorial nationalism which had been imported from the west. This nationalism was, in its very essence, anti-thetical to community life based on religion. But recently most of these countries are realising that, by trying to search for their identity in their national past in pre-Islamic days, they had actually lost their identity. Now they are returning to the religion. It is in this perspective that the resurgence of religion in Egypt should be viewed. Inspite of all the efforts of the late President Nasser, religion is once again going to play a very crucial role in the national life in the Egyptians. The author says ''It may be safely asserted that what has already happened in Egypt is going to happen in most other Muslim countries of the world.

ISLAMIC FUNDAMENTALISM, Causes, India

20. ENGINEER (Asghar Ali). Dilemmas of Muslim reformists; *mainstream*; 20, 23; 1982, February 6; 22-26.

"Reform movements could be of two types. First puritans who refuse to accept wider view of religion and confine it to revealed scriptures. The second category of reforms with which we are concerned is based on modern scientific, rational and humanitarian concepts. The motives for orthodox reforms are not always purely religious. Even Mahdi of Sudan, whom Arnold Toynbee describes as a 'zealot' launched his reform movement with manifest political motives, and same is the case with other movements. The above remarks throw much light on the recent fundamentalist movements in the Islamic world, specially the Islamic revolution of Iran. Unlike in North India there was an absence of orthodox reform movements (that is, reform movements of fundamentalist nature) in Maharashtra, as the Muslims in this region did not feel any immediate threat to their position. In recent times two reform movements, namely the movement for a common civil code launched by Hamid Dalwai and the movement for reforms in the Dawoodi Bohra community have attracted wide public attention. The author has critically evaluated the impact of these two movements in a great length and also described the causes of the failure of these reform movements.

21. ENGINEER (Asghar Ali). Islamic fundamentalism in India and its causes. *Link*; 25, 24; 1983, January 26; 24-26.

The author is of the opinion that the rise of Islamic fundamentalism is often talked about very loosely, especially in Indian context. "At the time of Moradabad riots an editor of a leading English daily even went to the extent of repeatedly writing articles and publishing letters to the effect that there was 'foreign hand' in the riot, meaning thereby that the Muslim fundamentalists from abroad were financing the riot". The Vishwa Hindu Parishad specially has made it a constant reference and has been playing up Meenakshipuram conversions out of all proportions and blaming them on the rise of Muslim fundamentalism". The phenomenon of fundamentalism would be different in different situation and for different people. While discussing the phenomenon of fundamentalism the author started with Jammat-e-Islami as it comes closet to the fundamentalist organizations and movements in the Islamic world. Untill recently the Jamaat was patronising the Student Islamic Movement (SIM) and it was considered as its student wing. The Jamaat maintained good relations with the regimes of the Gulf and Saudi Arabia. The Jamaat cadre is attracted towards the Khomeini regime as it is establishing the very theocratic state the Jamaat has been theoretically talking about all these years. The next organised group among Muslims is that of Jamiat-ul-Ulama. The other political parties and groups like Muslim League, Muslim Majlis, the Tamir-e-Millat etc. were also there but the League's approach is more pragmatic and empirical. The Shias of India (who do not have any separate political organization of their own) could be expected to be enamoured of the Khomeini kind of fundamentalism. However one does not see any noticeable reaction on their part. The author is of the opinion that there is no well organized reaction to Islamic fundamentalism among the Indian Muslims.

ISLAMIC FUNDAMENTALISM, Causes, IRAN

22. ELWELL-SUTTON (L P). Iranian revolution. *Int J*; 34, 3; 1979, Summer; 391-407.

In the present article the author has tried to offer an analysis of the religious situations as it prevailed in Iran on the eve of the religious movement of 1978. It may be that such an analysis will help to place Iranian Islam in correct perspective and to assess more accurately the part it plays in political and social development.

23. ENGINEER (Asghar Ali). Iran: A new interpretation of Islam. *Econ and Pol. Weekly*; 15, 40; 1980, October 4; 1654-55.

   While writing on Iran, western media does not keep in mind that Islam all through the medieval ages has been centred on the Legal Code known as Sharia where exploitation has no place. Recently Ali Shariati argued that the Shia Islam had been a challenge to the exploitative Umayyad and Abbasid empires but, since the Safavid period in Iran got itself feudalised, it was, therefore necessary to revive its early spirit by defeudalising it once again. According to Bani Sadr Islam allows owning of only that property which is created by one's labour. According to Payman it is the duty of militant Muslims to wage war against all those who are exploiters. Despite all this, no cohesive revolutionary theory is evolved and it requires an intellectual giant not the revolutionaries. Thus it is clear that in Iran there is a decisive struggle between the clergy and those radical groups fighting for the masses. The Mojahedin, although lacking the prestige of Khomeini, are certainly fighting for the poor and downtrodden, giving Islam a radical image. The masses may, if the Mojahedin chalk out their course cautiously, rally round them. Islam at last has a chance to show its revolutionary potentials.

ISLAMIC FUNDAMENTALISM, Causes, Middle East

24. SPRINGBORG (Robert). Islamic revivalism in the Middle East. *Current Aff Bull*; 56, 1; 1979; 14-15.

   Author says "for more than a decade there have been numerous indication that Islam might once again be on the march in the Middle East, but until the recent dramatic events in Iran. These signs have been discounted by observes and even participants in Middle Eastern politics". "Islamic revialism is sweeping through the Middle East in the wake of a failed Arab nationalism and Arab Marxism. Unlike the nationalists of an earlier era, however, the champions of Islamic revivalism are confronting elites who are not tarnished by association with colonial rulers and whose class origins do not place them in a tiny minority. As secularists many of these elites are isolated from the religious masses, but it is questionable whether distance in this one area is sufficient in and of itself for a new elite to grab the reins of power. Moreover, the authority of incumbent elites is already based on patch-work legitimation myths, so to include a slightly larger admixture of Islam in the ongoing search for legitimacy is not at all a difficult task. The plasticity of Middle Eastern politics and belief systems is such that a new movement having made its mark on the scene, will be accommodated alongside the old without causing a fundamental re-assessment of what may be contradictions only to an outside observer".

ISLAMIC FUNDAMENTALISM, China

25. ISRAELI (Raphael). Established Islam and marginal Islam in China from Eclecticism to Syncretism. *J. Eco Soc Hist Orient*; 21, 1; 1978; 99-109.

Author opine that Chinese Muslim's accomodating attitudes towards Confucianism (18th and 19th Century), far from creating a new syncretic religion, exposed and unbridgeability between Islam and the Confucian system. On both the spiritual and the material levels, the two cultures continued to survive side by side, separated from each other. The author suggests that the eclectic nature of the main stream of Chinese Islam was the primary reason why, in the 19th century, Islam could be revitalized, shrug off its Chinese accretions, and generate the vigorous wave of revivalism and ritual purity which restored to it its complete spiritual, and for a time even political, secession from the Chinese system. Thus, in contrast with the apologetic writings of previous periods, the late Ch'ing (1644-1912) and the beginning of the republican era (1912-on) saw a militant Islam, ready to confront both Confucianism and Christianity. There was, however, another kind of Islam, syncretic or marginal, which borrowed many enough Chinese characteristics to drift away from its source. This syncretic process gave birth to two different trends in marginal Islam in China: a. individual Muslims gradually foresook their tenets; b. Muslim messianic movements which were also alienated from "Established Islam". Paradoxically, it was marginal Islam of the messianic brand which gave to impetus to be whole movement of Muslim revivalism, which embraced established Islam as well. Had the Muslim rebellions in China succeeded, it is likely that marginal Islam, in its two varieties, would have found its way back to the mainstream of Islam.

26. ISRAELI (Raphael). Muslim revival in 19th Century China. *Studia Islamica*; 43; 1976; 119-138.

   After having treated the position of the Muslims in China during the late Ch'ing Dynasty (1644-1911), the author makes the following summarizing remarks: The use of Arabic and Islamic symbols was only the extreme manifestation of the Islamic revival in the 19th Century China. The mainstream of Chinese Islam (the "Old Sect") remained conservative, because the revival took place within the Chinese Muslim establishment which continued to seek a formula of accommodation of with the Chinese system. The co-existence deemed to be the only practical way of surviving as long as it remained unfeasible to overturn the government. Political readicalism was nevertheless, only part of the heightened Muslim awareness during Ch'ing dynesty, but it also signaled the unevitable result when Muslim political thought was taken to its logical conclusion: the creation of an independent Muslim state. The "New Sect" movement may therefore be seen as an avant-garde of Muslim revivalism in China and its purest expression since it divorced itself of practical considerations and sough to establish by force a political identity which other Muslims considered unobtainable. Although it is evident that the new sect drifted away from the mainstream of Chinese Islam theologically, apparently becoming associated with some forms of extreme Shi'ism, it is equally certain that it give Chinese Muslim political revivalism, its greatest impetus.

ISLAMIC FUNDAMENTALISM *compared with* ETHNICITY, India, South

27. BENSON (Janet). Politics and Muslim ethnicity in South India. *J Anthropo Res*; 39, 1; 1983; 42-59.

   A recent interpretation of Islamization among South Indian Muslims suggests that ethnicity may develop in response to internal needs to acquire status rather than as a result of interethnic competition. An alternative hypothesis, however, is that more emphasis on Islamic identity, including local ideas of orthography, occurs where Muslims face loss of power and status in a wider social system. This paper examines the structure of Muslim ethnicity and the process by which Muslim identity is maintained and intensified in the Telangana region of Andhra Pradesh. The historical role of Muslims in this area is compared and contrasted with that of Muslim in Tamilnadu, and it is argued that recent political changes have heightened Muslim insecurity in Andhra Pradesh and have resulted in a continuing emphasis on religious distinctiveness. It is further suggested that the concept of ethnicity is more useful than that of Islamization in explaining social change.

ISLAMIC FUNDAMENTALISM *compared with* REFORM MOVEMENTS, India

28. ENGINEER (Asghar Ali). Reform movements in Indian Islam. *Mainstream*; 19, 42; 1981, June 20; 16-20.

Like other religion Islam has also its vision which emphasizes on truth, justice, goodness, universal brotherhood, equality and so on. Koran explains that every action of man to perform religious ceremonies is not important. In fact it is faith of five pillars of Islam which is more important. First forty years of Islam are described as formative period of Islam, Later intellectual interaction left a deep imprint on both Islamic thought and practice which author call it the Mohayan Period of Islam. Author says that the main reason for the spread of Islam in India was sufis liberal attitude rather than Jurists and Ulema's rigidity. What, the author want to say that any reform or change has to be purposeful and in keeping with requirements of the community. Both revivalism and orthodox Ulemas are serving the cause of upper class. Thus in Indian context it has to be rejected and a new revolutionary theory should be negotiated for the better and change among downtrodden and backward Muslims. This ideology strives towards maximisation of social justice and human progress which is the real Islamic approach towards social progress.

ISLAMIC FUNDAMENTALISM *compared with* SECULARISM, Egypt

29. CRECELIUS (Daniel). Course of secularization in Modern

Egypt. *In* John L Esposito, *Ed. Islam and development*, New York, Syracuse University Press, 1980, p. 49-70.

This paper provides an excellent example of the complexity of change and the failure of accepted development theories to explain adequately modern Egypt's "retreat from secularism" in the twentieth century. It also traces the course of secularisation in Modern Egypt and suggests some reasons for the course that relationship between religion and the state have taken in revolutionary Egypt. Author feels it essential to review the relationship between religion and the state in traditional Islam. While concluding the article, author says that "one must wonder, therefore, whether the new Islamic revival is only a temporary phenomenon that will dictate only a momentary pause in the evolution of Islamic states and societies towards modernism and secularism or whether the religious movements will develop the strength to completely reverse the modernist-secularist trends to which they have always felt antipathy and do serious damage to the achievements of an earlier generation of liberal nationalists.

ISLAMIC FUNDAMENTALISM *compared with* ZIONISM

30. RAO (Madhav V). Islamic fundamentalism: Basic questions. *Mainstream*; 18, 46; 1980, July, 12; 21-22.

Much has been said about the revival of Islamic fundamentalism, particularly following the rise to power in Iran of Ayotallah Khomeini. Whatever the reason, there has been a curious diffidence on the part of scholars, commentators and understandably politicians to examine impartially the dynamics and implications of this phenomenon. The author says that "drawing a parallel between Islamic fundamentalism and Zionism is deliberate on my part, in as much as each of these has resulted in the assertion of the claim that a community organised along the lines of nation-state, Israel in the case of the Jews, and a variety of political entities ranging from Algeria to Indonesia in case of the Muslims, is in same sense responsible for the welfare of its co-religionist who happened to constitute a minority in other states. However, what is the proposition advanced by Khomeini now and implemented by Jinnah earlier? It is that a nation-state has to be based on theocracy and that its common law has therefore to reflect overwhelmingly the value system of the majority. The Islamic fundamentalism thus says that he not only has the right to be what he is in his own state. The reverse logic, however, does not apply in the case of communities that happen to be in minority with in the political confines of an Islamic fundamentalist state. Religion is a personal matter, and if there is one country that has throughout its tortuous history followed the principle of 'Live and let live', it is India.

ISLAMIC FUNDAMENTALISM, Concepts

31. ABDUL MOGHNI. Islamic mind of modern times. *Radiance*; 16, 34; 1981, January 4; 3.

Author says that "what is known as Islamic renaissance or resurgence of the Muslims has already taken place and is growing stronger every day". The rise of Islam is a fact of the day and is on the mind of every Muslim in whatever condition and region he be. The Muslim of India are as much affected by it as those of Saudi Arabia. It is actually Islam that has inspired the present generation of Muslims in all countries, who were otherwise losing heart, overwhelmed by the growing power of a hostile materialism. It is only the basic tenets of the ideology and system of life, presented by Islam, that the fundamentalists are out to propagate. The author further says that Islamic mind of modern times has cosmic vision and aims at a human character of the widest culture. The author says that "this mind is at work in India today, as much as in other countries of the world.'

32. SIDDIQI (Muhammad Nejatullah). Tawhid: The concept and the process. *Islamic Order*; 5, 1; 1983; 28-44.

Tawhid is the key concept in Islam. It sums up the Islamic way of life and present in a nutshell, the essence of the Islamic civilization. It is also the one term which describes the process of the Islamic transformation of an individual or a society". "The last half century in the world of Islam witnessed the emergence of powerful movements which sought to re-establish a direct contact between the Muslim mind and the Quran and gave it a fresh awareness of its larger role in human society". This is "the fact that humanity today is almost unanimously agreed on certain fundamental values for which Islam stands, such as freedom, equality, justice and democracy. While the Islamic movement has a lot to contribute by way of giving these values proper roots in the mind of man, it can also learn a lot from the variety of human experience in realising these values in actual life. In the end of the article the author says that "the contemporary relevance of these measures in the context of the process of tawhid is not difficult to discern. Should the contemporary Islamic movement stand equal to this task, it will be of great significance, not only for the world of Islam but for mankind in general. Should it remain pre-occupied with the revival of traditional religiosity in Muslim countries and involved in political struggle with their westernized elite, both the Muslims and the mankind in general will have to wait for a fresh attempt for the regeneration of humanity on the basis of tawhid".

33. TANZIL-UR-REHMAN. Islamic ideology. *Islamic Order*; 4, 3; 1982; 15-22.

In the present paper the author realize that the basic need of the Pakistan Muslim community was cohesion of the "fundamentalists" with the moderns. This paper emphasizes "that Islam is not only the name of certain beliefs, rituals and customs but a complete code of life and a system which governs all the fields of life. Authors says that "We cannot isolate politics or economics from all the different dimensions and aspects of organised life in Islam". "Islam does not confine its ideology or concept to the affairs of this world alone but it also encompasses the other world beyond this one, all, so to say, falling under its jurisdiction". "The basic and original sources of Islamic ideology are the Holy Quran and Sunnah". "Islam condemns every addition of permanent nature to beliefs, ritual and practices for which we find no support from the Quran and Sunnah". The author declares "We, the Muslims, with the grace of Allah, have the best ideology of the world. What is in fact needed is action integrated with the ideology. Unless the Islamic ideology motivates us in our actions other people will not be convinced of its utility nor have belief in its efficacy". "We have, however, to see and ask every one of us as to how we have applied and implemented Islamic ideology in various aspects of our life both individually and collectively".

34. ZIKRIA (Bashir A). Towards a Muslim intellectual renais-

sance. *Muslim Wld League J*; 9, 6; 1982, April; 26-28.

The glory of Islam in the middle ages was the glory that came from the cognizance of the real and its knowledge. The quran invites mankind to embrace Islam through reasoning and love, for this knowledge is required. That is why Quran says "Seek knowledge even if it be in China". It was the brilliance of Islamic minds, such as Ibn Rushd etc. who lifted the dark curtain of ignorance and fanaticism from Europe. The greatness of Islamic civilization was not power or army but was in the greatness of its minds, its great centres of knowledge and culture. And we see that decline of Islamic civilization started with the disappearance of these cultural centres. Need of the hour is to establish and revive such centres of knowledge and culture. In creating such an institutions however one's concern should not only be the scientific rediscovery of an uncremoniously buried past but also to rejuvenate a love and a thrust with in the Islamic world for the discovery of myriads of God's mysteries that are locked with in that macrocosm and this microcosm Allah created us as superior human being and its our duty to gain and work more and more for knowledge so that God may be happy with us.

ISLAMIC FUNDAMENTALISM, Critique

35. BAID (Samuel). How Islamic is Islamic resurgence. *IDSAJ*; 11, 4; 1979, April-June; 346-67.

    The author is of the view that "the so-called Islamic resurgence cannot be considered genuine as it lacks the moral approval of large section of people in Muslim countries. The forces behind this phenomenon have their own vested interests." "Saudi Arabia and Libya, the two fundamentalist Muslim countries, desirous of maintaining their religious hegemony on the Muslim world, helped these movements". Different Muslim countries has their different vested interests which they want to achieve through this Islamic resurgence or through the exploitation of Islam.

36. ENGINEER (Asghar Ali). Islamic world: Myth and reality. *Mainstream*; 18, 42; 1980, June, 14; 20-22.

    The author says that "terms like Islamic revivalism, Islamic fundamentalism etc. being used frequently to describe hap-

penings in the Islamic world are loose and inadequate".
"The western media, one can therefore safely conclude, are
deliberately giving a religious orientation to events in the
Islamic world". Islamic countries are economically and
technologically backward. Many Arab countries still have
a large segment of Bedouin population. The Bedouin led
groups are opposed to modern technology and wanted a
crusade against corruption in the ruling Saudi family. To
understand the current upheaval in the Islamic world it is
necessary to have both a sociological and a historical view
of the situation. In the Islamic world too one finds today
various groups at different stages of historical development. The ruling classes in the Islamic countries continue
to be feudal and conqueror in character. No wonder that
none of these countries, from Pakistan to Saudi Arabia has
been able to evolve a democratic polity. Here it may be
noted that what is sought to be revived is not the true spirit
of Islam or fundamental Islam as is often claimed, for there
is nothing like pure or fundamental Islam. Every one practised and tries to understand Islam through his own spatio-
temporal frame and calls it pure Islam. In case of Iran
revolution "it is a massive protest movement against a
politically repressive regime as well as against social
change and not pure and simple religious movement, as is
often claimed. As against Iran, the revolutionary change in
Afghanistan was the result of a military coup. The author
says that "revival of fundamental Islam" is a myth created
by powerful vested interests from the Islamic world as well
as from the western world in order to divert people's attention from the concrete tasks of solving economic
problems.

37. JAIN (Girilal). Power of Islamic revivalism: Change seen as
threat to identity. *Times of India*: 1979, November 7; 8; 8.

Stressed that secular minded liberal Muslims are greatly
embarrassed by the actions of Ayatollah Komeini and Ziaul
Haq. They are also making an effort to convince their fellow
religionist that it is not necessary for them to follow the
fundamentalist's path. Such an attitude will be set back for
the fundamentalist in entire Muslim world. It is fashionable
to describe the triumph of the Ayotallah in Iran. The fundamentalist had, however, begun to assert themselves suc-
cessfully in Pakistan. The author also discussed the relative
weakness of the revivalist sentiment in Pakistan.

38. MEHROTRA (O N). Resurrection of Islamic fundamentalism. *Strategic Analysis*; 3, 3; 1979, June; 123-27.

    Author is of the view that "in substance, Islamic fundamentalism as propagated and understood cannot be implemented in a modern state. There is bound to be seperation of powers between temporal and spiritual authorities. Nationalism and modern education will not allow adoption of the old Islamic state system. Other ideologies, such as capitalism and communism, may also prevent the spread of Islamic fundamentalism. All Muslims are fundamentalists but this does not mean that they are for adopting a system which was prevailing in the seventh century, and has not much relevance in the present day. This realisation can be achieved with the help of enlightened Muslim leadership and public opinion. It is generally believed that resurrection of Islamic fundamentalism will met with only partial success."

39. MOIN SHAKIR. Politics of "Islamic fundamentalism" *Mainstream*; 18 42; 1980, January 14; 15-20.

    The overthrow of monarchy in Iran, Ziaul Haq's pledge of establishing Nizam-e-Mustafa and the world wide condemnation by the Muslims of the attack on Kaaba Mosque, etc., are sited by journalists and commentators as example of Islamic solidarity and spiritual unity which signify 'the return of Islam', 'New wave of Islam', 'Islamic Revolution', 'Islamic resurgence', 'Islamic search for identity' and what not. An attempt is being made to interpret the life of a people wholly from the standpoint of religion. This is bound to lead to superficial conclusions. In this paper an attempt has been made to suggest that it will be a serious error to describe the problem and responses of the Muslim peoples in terms of religion and ethos of Islam. The nature of Muslim politics can be seen only in terms of the economic realities of Muslim societies rather than the religion of these societies. It will be wrong to emphasise the 'universal' and 'central' essence of the recent Islamic resurgence. The content of this 'fundamentalism' varies from country to country. To conclude, Islamic fundamentalism has been used by nondemocratic rulers for the legimation of their authority. Otherwise it will be extremely difficult to justify military regime of Zia or the Saudi kingship or Islam—minus democracy notion of Khomeini on the grounds of Muslim religion. Secondly Islamic fundamentalism serves as an ideology to the exploiting ruling classes. Thirdly, Islamic fundamentalism in different Muslim countries has been turned into a design to destabilise the liberal and pseudo-democratic countries. Forthly, the Muslim countries, even if they have liberal or democratic set up cannot affectively counter the threat of fundamentalism unless radical changes are made in the economic system. Lastly, the salvation of the Muslims all over the world lies in the struggle against exploitation and in equalities. The struggle should aim at the establishment of a society on the basis of the highest values of humanity.

40. NOORANI (A G). Islamic fundamentalism. *Indian Express*; ʾ979, November 1.

Author says that the Prophet and his Caliphs did not devise any model of "an Islamic polity". Therefore, the endeavours of the "fundamentalist" to draw institutional forms from ethical precepts are wrong and futile. To prove this thesis, the author has advanced all his evidences and arguments and proceeds to condemn what he calls "the quest for an Islamic constitution", obviously meant for an Islamic state in the present day world. The author puts a question: Why an Islamic State? He also raised certain questions of jurisprudence also, to justify his thesis of modernising Islam.

41. WARIAVWALLA (Bharat). Islam's new postures. *Seminar* (Annual Number); 245; 1980, Jan; 60-63.

Distortion of history is quite common now a days and specially in India. This is general feeling among Hindus that Muslim were communal for Hindus when they ruled over India. At present India is a secular state which divorce the affairs of state from religion. Main factors in Zia's rule in Pakistan and Khomeini's regime in Iran are not merely socio-economic. Specially in Iran we cannot ignore the relevance of Islam to the politics. Many leaders in Muslim world cynically manipulated religious symbols in order to create modern nation-states. The question arises that why it is that a liberal democracy has just not sunk any roots in the Muslim countries. Indian Muslims are divided along, class, caste and sect lines. The revival of the faith is not likely to lead to any greater unity than exists today among Islamic countries. Islam has seldom cemented political fissures. In recent years Pan-Arabism of sorts has vaguely succeeded in transcending national difference like Libya's Pan-Arabism has been a non-starter. The Islamic world is not in any less of a dissarray as result of the resurgence of faith. Muslim States continue to querrel among themselves as they have in the past — Algeria and Morocco, the Yemens, Syria and Lebanon, Pakistan, Afghanistan, Iran and Iraq — the list is long.

ISLAMIC FUNDAMENTALISM, Economics

42. ABU SAUD (Mahmud). Money, Interest and Qirad. *In* Khurshid Ahmad, *Ed. Studies in Islamic economics*, Leicester, Islamic Foundation, 1980, p. 59-84.

Author warns that "it is dangerous to adapt an economic system that does not emanate from and correspond with the ideology, in fact such a trial is doomed to be a failure. Thus there is no "Islamic Economics" unless there is an Islamic ideology prevailing and applied in a Muslim community." The author has tried to explain about money, Interest, and Mudarabah or Qirad in terms of Islamic ideology and says once the meaning and nature of these terms have been clarified, the study of "Islamic Economics" will be easier to follow. The author gives special treatment and attention to qirad in this article under various headings such as Qirad and Banking; Qirad under the present Banking system; Qirad in a special Banking system; Qirad and Interest. He also discussed about Zakat in length.

43. ALI ASHRAF. 'Islamic' economic thinking at cross-roads. *Mainstream* (Republic Day Number); 1983; 35-39.

According to the author "In recent years, in the context of what has come to be known as Muslim fundamentalism, considerable literature has grown up on so called Islamic Economics and Economic System of Islam." "Contemporary Islamic resurgence is not simply an exercise in political activism... (but) a move towards rediscovering Islam as the basis for the new social system ... and embraces almost all aspects of their life, intellectual, social, political, economics, educational, cultural and international". (Prof Khurshid Ahmad) A number of writers draw upon the past to show the historical roots of Muslim economic thinking. Not with standing the fundamentalism and revivalist trappings and the claim to be free from "political activism" the factors working in the emergence of contemporary 'Muslim' economic thought are very much contemporary and not unrelated to modern economic and political realities. Nijatullah Siddiqi admits that Muslim economic thinking is a twentieth century phenomenon. The founders of present-day Islamic fundamentalist movements, Sayyid Qutb, and Mawdudi, both stood for a *Laissez faire* economy. Various author consider that if the twin measure of prohibition of interest and payment of *Zakat* are faithfully implemented by the Islamic collective, together with the Islamic law of inheritance, there will be no ground left for capitalism to grow. In modern Islamic fundamentalism, however there also exist other, more radical trends. Whatever it be, it indicate serious fissures in modern Islamic fundamentalism. A number of Muslim countries adapted socialism as the objective of their state policy. The struggle of the people may have a long way to go, through zig zags and ups and downs. But they are hardly likely to allow the moulding of Islam according to capitalism.

44. CHOUDHURY (Masudul Alam). Principles of Islamic economics. *Islamic Stud*; 21, 2; 1982, Summer; 89-107.

Author's objective is "to delineate in nontechnical economic language the principles of Islamic economics in as far as they constitute the philosophical basis of this economic system. Discussed are: A. Principles of Islamic economics: 1. the principle of tawhid and brotherhood, 2. the principle of work and productivity, 3. the principle of distributional equity. B. Analytic implications of the Islamic economic principles as mentioned under A.C. Policy basis of the Islamic economic principles. 1. Abolition of Riba, 2. Institution of Mudarabah, 3. Abolition of Israf, 4. Institution of Zakah. The author concludes by recommending that Muslim nations "should clearly identify their national priorities in social values and link them soundly with their economic development strategies".

45. ENGINEER (Asghar Ali). Islamic economics: A close look at Bani Sadr's theory. *Mainstream*; 19, 27; 1981, March, 7; 22-26.

Author says "what is happening today in Islamic world needs to be seriously analysed in the light of structural changes occuring in the socio-economic set up of most of the Muslim countries. The upheaval in Iran has to be understood in the light of socio-economic forces generated as a result of the oil boom. Dr. Ali Shariati and Bani Sadr are eminant Islamic revolutionaries who have attempted to propound a self consistent theory of Islamic revolution by re-interpreting the teachings of the Koran, the Prophet's Sunna. In this article the author discussed some aspects of Bani Sadr's views on Islamic economic propounded by him in his book *Iqtisa-i-Tawhidi*. He is critical of communism in certain respects. Bani Sadr feels that in Islam the right to property is not absolute. He distinguishes between ownership based on force (Malkiyat-i-zor) and ownership based on work (Milkiyat-i-Khususi). He holds that ownership in a capitalist society is based on force and is therefore un-Islamic. Bani Sadr maintains that nationalisation is not only approved by Islam, but it is desirable. Bani Sadr makes it very clear that the real aim of an Islamic society is to liberate man. Marx too desires to see man liberated from economic bondage. Bani Sadr is fully aware of the appeal of secular ideologies, like Marxism and hence feels it necessary to reinterpret Islam in such a way as to ensure socio-economic justice. Throughout the book Bani Sadr unlike the upholders of the revivalist trends, tries to give utmost prominence to the elements of economic justice in Islam. The revivalist on the other hand play down the concept of economic justice and, instead, emphasise *ibadat* and *uqubat*. It is repeatedly emphasised in the book, and rightly so that socio-economic justice is the fundamental principle of Islam. Bani Sadr says that Islam has prohibited interest in all its forms. Bani Sadr is one of those few Islamicists who have tried to present a very different image of Islam in the light of modern development.

46. FAROOQI (Abdul Hai). Economics in the Islamic framework. *In* Abdul Hameed, *Ed. Islam at a glance*, New Delhi, Indian Institute of Islamic Studies, 1981, p. 101-109.

Islam provides a code of conduct—doing things which the Quran and the Sunnah ask him to do and refrain from what they prohibit, points out the author. Islam also prescribes its own economic order which refers to an economy, the basic structure of which will be laid down in accordance with the injunctions of the Quran and the Sunnah. Islam does not grant economics and independent status. It requires that all forces inherent in man and all his economic activities must be directed to Divine Pleasure. Author also discussed about the "interest free economy and Zakat." He says that "moderation in consumption is an important characteristic of Islamic economy. Moral persuasion will have an appreciable role in an Islamic banking system, says author. In the conclusion, the author suggest "the Islamic economic system will have to operate with its own tools. The prime basis will be participation in income and risk sharing. The economic goals will be those that have been defined by the Quran and the Sunnah. All economic policies will have to be so directed as to achieve the welfare of people and the pleasure of God".

47. GHOUSE (Agha Mohammad). Role of work, employment and money under Islamic principles: A basis for new national and International Economic Order. *Islamic Order*; 4, 2; 1982; 73, 82.

Discussed the role of work, employment and money as the basis for new national and International economic order under Islamic principle. The whole article is divided under the following headings which will also give an idea about the article. World poverty and growth—Islam's socio-economic approach—Work and employment in Islam—Money expenditure and investment— Profit-sharing investment and Banking.

48. KAHF (Monzer). Contribution to the theory of consumer behaviour in an Islamic society. *In* Khurshid Ahmad, *Ed. Studies in Islamic Economics*, Leicester, Islamic Foundation, 1980, p. 19-36.

The principal focus of this paper will be on the impulses that motivate and the goods that are pursued by the consumer in his choice among goods and services contained in the attainable set, and on the macro economics of this behaviour, given the aximatic system of Islam. Whole article is divided under five sections. "Section I will briefly highlight the ideas on consumer behaviour developed under the capitalist and the Marxist systems. Section II will analyse consumer rationality under the Islamic axioms. Section III will be devoted to the consumer's decision to allocate his income between saving and final spending. Section IV will derive the macromodel of income allocation in an

Islamic society. The fifth and the last section will summarise the findings and the conclusions."

49. al-MAHDI (Sadiq). Economic system of Islam. *In* Salem Azzam, *Ed. Islam and contemporary society*, London, Longman, 1982, p. 101-19.

Describes the economic aspects that were practised by the regime at Medina, in this paper. "But certain factors conceived in the womb of Islamic society destroyed the system and proceeded to develop a hybrid economic system which is Islamic only in the loose sense of belonging to Islamic civilization". The author further says "our ancestors have bequeathed us a rich corpus of ideas and regulations which is most relevant for a comprehensive understanding of the economic system of Islam". The author also discusses about interest, private ownership, zakah and about the state intervention. About the economic system of Islam, the author says, even after 1400 years the system is as fresh as ever and as relevant as ever.

50. MOHAMMAD NAIMUDDIN, Note on contemporary thinking on Islamic economics. *In* Anwar Moazzam, *Ed. Islam and contemporary Muslim World*. New Delhi, Light and Life, 1981, p. 107-17.

As large part of ideological conflict revolves round the economic problems, attempts are being made to evaluate the economic problems from Islamic view point. A lot of literature has been produced in the name of "Islamic Economics". The object of this paper is to present briefly and evaluate critically representative Muslim thinking on economic systems, specific economic issues and on policy instruments, in the Indian sub-continent. According to Abul Ala Maududi, Islamic solution to the economic problem lies in 1. The differentiation of *Halal* and *Haram*, 2. Observance of Islamic trade code 3. Discouragement of luxurious living, 4. Prohibition of interest, 5. Payment of *Zakat*, 6. Basing the whole economic system on the moral principles of Islam. The approach of Islamic economists is rather conservative. There is tendency to shy away from the problems that are difficult to deal within the orthodox frame. The question that arises in one's mind as one goes through the literature on Islamic economics is whether the Islamic economy will be fundamentally different from other economic systems in its goals, institutions and techniques. Perhaps the answer is in the negative. It appears that the Islamic economy will differ from a capitalistic economy in the absence of interest as a source of income, and from a full-fledged socialist economy in its permission of the institution of private property. The author says "Whether the Islamic economy will be nearer to a socialist economy or a capitalist economy will depend on the value judgements of those who will prepare the blueprint of an Islamic economy.

51. NAQVI (Syed Nawab Haider). Ethical foundations of Islamic Economics. *Islamic Stud*; 17, 2; 1978; 105-136.

Describes that "excessive materialism (of the West) in the end proved to be self-distructive. There has occurred, as a result, a remarkable resurgance of interest throughout the Muslim world in the Islamic way of life. Economic prosperity among the oil-rich Muslim countries has forced even the Western Countries to take Islam seriously, if only for transitory potential ends". Author suggested that "the first step towards understanding the economic principles suitable to the Islamic perspective on life is to understand clearly its ethical system". "Unity, equilibrium, human freedom and responsibility provide such a set. After applying this criterion set to the existing economic systems of socialism and capitalism we find that these systems do not seem to be Islamic since they violate almost all the ethical exioms". The author says "this also confirms our hypothesis that there exists an imperative need to develop fully Islamic economic system, because it alone can truly satisfy all the fundamental ethical axioms"

52. NAQVI (Syed Nawab Haider). Islamic Economic System: Fundamental Issues. *Islamic Stud*; 16, 4; 1977; 327-46.

Author starts by admitting that "we do not have at present any thing like Islamic economics in the sense of a body of mutually consistent doctrines of sufficient generality, simulating a real Islamic society. The author's axioms are: a. Islam assigns a central place to the individual in the universe; b. however, this freedom entails "responsibility" which in turn implics ethical (and social) constraints on his "natural" freedom. Accepting these axioms, he then deduces the "rules" or "principles" for a typical Islamic economy (choice of appropriate economic frame work; basic policy objective; choice of specific economic policies). Concluding, the author remarks that there is no urgency at all to invent a whole economic system in a hurry. Any attempts to artificially telescope the evolutionary process involved in framing rules of the game in a real-life Islamic Society, is fraught with great dangers. "To those who stand outside the cycle of beliefs and passions, which go into the making of political debates, only scientific objectivity can provide the right, or most nearly right answers to the challenge posed by the need for establishing in Pakistan an economic order based on the teachings of Islam.

ISLAMIC FUNDAMENTALISM, ECONOMICS compared with Western

53. WILSON (Rodney). Economic change and re-interpretation of Islamic Social Values. *Brit Society Mid East Stud*; 9, 2; 1982, 107-13.

Examines Islamic economic ideas from a western perspective. He focuses on what he calls three "assumptions in all western neo-classical economics": 1. Prices are the major means of allocating resources; 2. Present consumption is preferred to future consumption; 3. Individuals are risk averse. He illustrates their relevance/irrelevance in an Islamic context. Dealing with the application of Quaranic views on usury, he indicates some ways in which the strictures on usury are interpreted in the Muslim world today: Islamic banking (providing for profit sharing) and direct prohibition of interest. As for insurance: In Islam there are two objections to insurance: 1. A fundamental one (fatalistic) and 2. With respect to the way insurance companies invest their funds. Is there an Islamic economic system The author quotes Maxime Rodinson (Islam and Capitalism): "The correlation between Islam and any particular economic system has emerged as being very largely inconclusive ..." but declares himself "to try to reach a conclusion with respect to a whole system, as Rodinson attempts, is certainly an over-ambitious task, and one which in my views is doomed to failure".

ISLAMIC FUNDAMENTALISM, ECONOMICS, CONFERENCES

54. FINNEGAN (J M). Restoration of Islamic law Vs. Secularization. The first world congress on Islamic economy. *Cemen Reports*; 4; 1976 publ, 1978; 25-140.

The First World Congress for Islamic economy, organized by the faculty of the university 'Abd al-Aziz of Jeddah, was held in mecca 21st-26th Feb. 1976. "grouping some 350 specialists, theologians, canonists and economists from 33 countries of the Islamic world and such places as England, India, Japan and U.S.A. the Congress was considered an excellent opportunity for the Islamic world and such places as England, India, Japan and U.S.A, the Congress was considered an excellent opportunity for the Islamic world to take stock of itself and define its basic attitude on economic matters in their relation to man's destiny". The Congress proposed to offer a general scientific exposition of Islamic economic theory. The article focuses on: the religious character of the Congress, main points in the discussion; private property and inheritance, Zakat, legitimate acquisition of wealth, an Islamic Bank, and insurance. To the observer "it is plain that the Islamic world vision thus outlined continues to consider Muslim society as a "closed" structure With the exception of Indian delegate" ... "no other delegate seems to have taken up the problem of non-Muslim minorities in Muslim states nor the more general question of the multiple relations, economic, cultural and political, between states. The dominant attitude is one of seeing "the whole weight of Islamic economy organised in view of Islam's independence and progress. The suspicion which still clings to the notion of a natural law in so far as it would offer a common ground for the creation of a common society irrespective of religious diversity has not yet been dissipated". Added are the final recommendations of the conference.

ISLAMIC FUNDAMENTALISM, Economics, Development

55. AMEER ALI (Shaikh). Islam: An alternative approach to economic development. *Muslim Wld League J*; 9, 2; 1981, December 6-9.

While looking at Muslim countries in the world, the political, economic and military upheavils and the popularity of revivalism the main question arises whether Islam is a viable alternative to the existing model of development, what are its characteristics and what are its problems? From two models, socialist and capitalistic Islamic alternative differs from all these models. Ethics and morality as set by the Quran and the Sunnah are the most basic determinants of Muslims behaviour where it is political economic or otherwise. For the economic path of society---Islamic model some what adopts the middle way followed by the mixed economics but with an important differences of toning the market forces and testing the sharpness of objective reasoning with the grind-stone of the Shariah laws. To conclude author says that the Shariah and the economy can test each other's validity or practically and rectify their shortcomings by using the weapon of *Ijtihad* or reasoning sanctioned by Shariah itself. The development path along with an Islamic economy proceeds should exhibit three of the most essential qualities like justice, benevolence and moderation.

56. CUMMINGS (John Thomas) and others. Islam and modern economic change. *In* John L Esposito, *Ed. Islam and development,* New York, Syracuse Univ Pr, 1980, P. 25-47.

Religion is not only related to politics but also is integral to the economic structure of an Islamic state. Oil wealth, as well as the political resurgence of Islam, have reawakened interest in the potential impact of Islam on economic development. John Thomas Cummings, Hossein Askari, and Ahmad Mustafa in this article focus on the resources of traditional Islamic economics, its compatibility with western capitalist thought, and its relationship to current socioeconomic change in the Middle East. Islam addresses itself to many aspects of economic development --- private ownership, taxation, interest, income distribution, etc. The authors maintain that Islamic principles of economics do not necessarily preclude rapid economic growth. On the contrary, Islamic principles advocate factors which are generally regarded as essential to economic progress --- private property, profit incentive, hard work and eternal reward for economic success.

57. ISHAQUE (Khalid M). Islamic approach to economic development. *In* John L Esposito, *Ed. Voices of resurgent Islam,* New York, Oxford University Press, 1983, p. 268-76.

A major theme of the Islamic resurgence is economic/social justice. Poverty, illiteracy and distribution of wealth have continued to plague most Muslim societies. Western economic system and Arab socialism have not altered the situation appreciably. Muslim revivalist attribute this failure to continued reliance on imported economic models. Instead, an Islamic solution to economic development is advocated --- one which is rooted in the beliefs, values, and experience of the people and is, therefore, better able to inspire, motivate, and assure Islamic social justice. In the present paper the author summarizes the critique of western economic models and presents an Islamic social framework for economic development.

58. KHURSHID AHMAD. Economic development in an Islamic framework. *In* Khurshid Ahmad and Zafar Ishaq Ansari, *Ed. Islamic Perspectives*, Leicester, Islamic Foundation, 1979; p. 223-40.

"A major challenge confronts the world of Islam: the challenge of reconstructing its economy in a way that is commensurate with its world role, ideological, political and economical. The author in this article has tried to see whether the Muslim world is clear about this fundamental question or not. The author has stressed for the need to clearly identify the Islamic ideal of economic development to measure the distance between this ideal and the present day reality of the Muslim world in such a way that an Islamic framework of life may ultimately be evolved. This formulation of the problem has immediate relevance for the Muslim economist and planner. The author suggested that "it would be only through sustained research by a team of economists, by unceasing original thinking and, above all, by a great deal of practical experimentation that we might be able to discover an Islamic road to economic development".

59. MIR KHAN (M). New National and International economic order on the basis of Islamic concepts. *Islamic Order*; 4, 1; 1982; 91-99.

In this study an attempt has been made "to understand what is being called the "New International Economic Order" and to see if conformity exist between Islamic principles and proposals for such a new order in international and national economic affairs. For economic relations among nations the generality of proposals evolved and advocated by the developing countries conform with sound Islamic ethics ordaining the rich to assist and cooperate with the less rich and the poor. For redressing national economies on completely Islamic principles, each national is obliged to take account of its special circumstances of the composing faiths of its population, its existing economic structure and its problems in production, trade and monetary system. The ruling spirit in Islamisation of its economy to the extent possible should be the just equitable participation of its population in national production and its distribution".

60. ZARABOZO (Jamal-al-Din). Islam and economic development. *Muslim Wld League J*; 11, 2; 1983, Nov.-Dec.; 35-37.

Author says making the (normative) decision to achieve economic development at the expense of the values of the society is indeed injurious to the society as it was done in Tunisia by abolishing fast as a hinderance to economic growth. Islam does realise the necessity of meeting people's basic needs. But values are more high than mere economic growth. Value of Islam describes their own definition and path of economic development and that the economic problem of man is solved with in the Islamic framework for society. Zakat which is most important economic factor in the Muslim society is not practiced even in Muslim countries. This may be because of international economic system. To conclude author says that our Muslim nation of the world have shown no signs of change. Let us hope that Muslim world will stop looking else where for answers and truly seek to attain development as it is defined by their religion, Islam.

ISLAMIC FUNDAMENTALISM, Economics *in relation to* Politics

61. GILANI (Ijaz Shafi). Political context of Islamic economics: High and low road strategies. *In* Khurshid Ahmad, *Ed. Studies in Islamic Economics*, Leicester, Islamic Foundation, 1980, p. 131-42.

One of the primary functions of any science is classification. Whether in natural or social sciences, the task of an analyst is to classify disparate events under formal classes. Here also, the author has categorise the discussion under three classes: Theory, policy and practice for the purposes of a discussion on the political context of Islamic economics. The paper is an attempt to perform a two-fold task of A. Clarifying the distinction between theory, policy and practice, B. Focussing on the policy aspect of Islamic economics. While concluding this paper, the author points out that the "Muslim economists are making an effort to work out a theory and a policy for Islamic economics."

ISLAMIC FUNDAMENTALISM, economics *in relation to* Social Welfare

62. ZARQA (Anas). Islamic economics: An approach to human welfare. *In* khurshid Ahmad *Ed. Studies in Islamic economics.* Leicester, Islamic Foundation, 1980, p. 3-18.

The author has made an effort, in this paper to shed some light on certain methodological and philosophic aspects of Islamic economics and to illustrate its unique approach by reflecting upon an Islamic social welfare function. The author says "I shall argue in this part and elsewhere in this paper that distinct and meaningful Islamic economics is possible, nay, necessary. Briefly, my argument is that economics is not as innocent of value judgement as we are often led to believe, nor can it ever be. Neither is Islam devoid of positive assertions about economic reality. We can then replace on Islamic value judgements by Islamic ones, and add to the economist's stock of positive assertions. Islamic assertions, then work out the consequences.

ISLAMIC FUNDAMENTALISM, Economics, interest

63. al-JARHI (Mabid Ali Muhamed Mahmoud). Relative efficiency of interest-free monetary economics. The fiat money case. *In* Khurshid Ahmad, *Ed. Studies in Islamic economics.* Leicester, Islamic Foundation, 1980, p. 85-118.

The purpose of this paper is to challenge the traditional institutional arrangement of paying interest on money as an efficient monetary policy. The author also concentrated on fiat means of exchange. The paper introduces a set of fiat means of exchange into an economy and a few related questions. The conclusion of this paper make it obvious that economics with no interest payments on borrowing and no bank multiple creation of money are most optional between the different institutional arrangements considered. This means that it is most efficient if the government initially provides its own money free, lends it free and imposes a 100% reserve ratio on banks. The author says "we expect such conclusions to be more of a surprise to economists in Muslim countries, than to the young economist of the western world. Author hope that "our conclusions will be an inspiration to the economists of the Muslim world to revamp the economies of their own countries and rid them of the traditions of old western economics. The author hopes that this will be a first step towards constructing an Islamic economic theory.

64. MUHAMMAD SAMIULLAH. Prohibition of Riba (interest) and insurance in the light of Islam. *Islamic Stud*; 21, 2; 1982; 53-76.

"Islam prohibits all transactions involving interest. Interest is neither trade nor profit. It is a means of exploitation and concentration of wealth" ... insurance societies of the modern world are not in accord with the spirit of Islamic law for the reason that they operate on the principle of fixed premiums and fixed policies. This is the same principle of predetermination of risk as is followed by stock companies and it stands disapproved in Islam because of its speculative nature. Pre mutual institutions accord with the spirit of Islamic law. The article discusses more in detail these prohibitions paying attention to the basic notions of the Shariah or Islamic law (which are "ethical in character") relating to transactions and gives an alternative solution.

65. NOORZOY (M Siddieq). Islamic laws on Riba (interest) and their economic implications. *Int J Mid East Stud*; 14,1; 1982; 3-17.

The word *riba* means increase which corresponds to the word interest. In both cases the increase refers to the amount beyond what is owed. Thus, the strictest interpretation that can be given to the word *riba* is that it means interest — an amount, or rate, due above the principal of a loan. The purpose of this paper is a limited inquiry about some aspects of the controversy on *riba* as they affect contemporary economic transactions in a Muslim society.

"Some of the issues raised in this context clearly also have relevance to a larger model which encompasses economic transaction between a fundamentalist Muslim state that adopts the full working of an interest free economic system and the non-Muslim world which functions with a positive rate of interest. The analogy would also apply between Muslim state at large and the rest of the world if the former invoke Islamic laws on *riba* along the same lines".

66. QADRI (S.M.). Quranic approach to the problem of interest, in the context of Islamic social system. *Islamic Cult*; 55, 1; 1981, Jan; 35-47.

It is stressed that *riba*, as a Quaranic term, should be studied in the context of the Islamic social system as envisaged by the Quran and al-Sunna. The Quran does not confine "economic life" to the mundane life of the people (integrity of human life). With this basic point in mind, the author deals with the Quranic data relating to the premises of the economic system, particularly the prohibition of taking interest (*riba*). It is concluded that today the time appears to be very opportune for the development of social science on Islamic grounds, be it without the need of discarding the western economic institution in toto. "What we need is the Islamization of western economic institutions. For instance, by the elimination of the element of interest we can adopt the banking and cooperative system in our life. Instead of paying interest the depositors can be paid dividends as share-holders and dividends are the best alternative, to interest". The author argues for the contribution by Muslims of a perfect social system to what he calls the material western world "which is completely devoid of spiritual elements of sublime life". Such a task demands the "establishment of an international academy of Islamic social sciences".

67. ZIAUDDIN AHMAD. Qur'anic theory of Riba. *Islamic Quart*; 20, 21, 22, 1-2; 1978; 3-14.

Examines Riba al-Jahiliyya (Riba: interest during the pre-Islamic days) and compares it with present day bank interest. It appears that *Riba al-J* was an atrocious kind of transaction which resulted in heavy indebtedness; in modern times banking is organised by powerful institutions. The need of imposing a check on economic exploitation is felt every where. The resemblance between *Riba* and present-day bank interest is still a source of confusion for the Muslims who want to organise their economy strictly on the principles of Islam. Many a Muslim thinker, both modernist and traditionist, has tried to solve this problem. "The discussion is still continuing and this shows that the problem requires a thorough investigation especially in the context all other socio-economic issues of the Muslim umma. However, it is felt that in order to eliminate the chances of economic exploitation ways and means should be devised to bring the present bank closer to the notion of *Bay* (Bay also involves increase but it is due to transaction itself, not due to the time factor) which is opposed to *Riba* in the Quranic system". The author suggests that the method of sharing in the profit and loss between the banks and depositors may remove the resemblance between *Riba* and bank interest.

68. ZIAUL HAQUE. Nature of Riba al-Nasi'a and Riba al-Fadl. *Islamic Stud*; 21, 4; 1982; 19-38.

Concludes historical analysis of *riba* (interpreted in various ways: sometimes as usury and simple rate of interest, and sometimes as an unearned income) as follows: "The general hypothesis of some modern Muslim scholars that *riba* in the modern industrial economy, inherit only in money-leander's capital, must be re-examined in the light of the jurist's interpretation of *riba*. The modern production process is a highly complex organization which is closely related to and dependent on the given structure of distribution of wealth. The modern science of economics take interest of capital as a part of the over-all problem of distribution, i.e. how national wealth is distributed among various factors of production. The crucial problem for "Islamic economics" is to re-define and re-interpret the medieval concept of *riba* in the modern mixed economy. To term moneylender's interest as *riba* and abandon all other economic areas and sectors to rapacious profit motive and self interest as a one sided interpretation".

ISLAMIC FUNDAMENTALISM, Economics, Islamic Banking

69. MUHAMMAD UZAIR. Some conceptual and practical

aspects of interest-free banking. *Islamic Stud*; 15, 4; 1976; 247-67.

Banking is an important financial intermediary and vital institution in the economic structure of any country. It mobilises savings and idle funds in an economy and makes them available to those who can make better and fuller use of them. The author points out towards "an important question that agitates the mind of those who think on Islamic economics, whether proponents or critics, is the feasibility of interest-free banking. The author in this article, has tried to discuss here, briefly, certain conceptual and practical aspects of the issue.

70. NIENHAUS (Volker). Islamic Banks --- a short survey. *Islam Mod Age*; 14, 1; 1983; 1-10.

"At present, most of the activities for the establishment of new Islamic banks is to be found in Muslim countries of non-Arab, Africa and South East Asia". Since the late 1940s there have been discussions, especially in Pakistan, on interest free Islamic banking. "The initiator of the first practical and successful experiment in Islamic banking in Egypt in the 1960s, however, followed a quite different line of reasoning" (his model being the German saving banks). The author concludes his survey as follows: a much greater contribution (than recycling activities on Islamic commercial banks) to the development of non-oil countries of the Muslim world" could spring from the realisation of the Islamic saving bank idea. This type of banking should have priority in the next years, although in the long run probably all types of Islamic banks could perform specific and complementary tasks in an emerging Islamic financial system. For the time being, Islamic Bank should be seen first of all as an instrument for the promotion of economic development and the economic cooperation within and integration of the Muslim world but not as an instrument for the complete and rapid elimination of interest from the economy.

71. SIDDIQI (Muhammad Najatullah). Banking in Islamic frame-work. *Islam Mod Age*; 8, 4; 1977; 5-20.

The basic reform advocated by the author is replacement of interest by profit-sharing as the basis of bank's advances. It would take care of "the evil consequences of the system" as noted by the writer. "One further advantage is to ensure active involvement of the banks in the productive ventures financed by them. Bank's income being derived directly from the actual profits realised in the ventures they finance, their expertise will also be focussed on efficient decision making and proper management in these ventures". The author further traces the impact of interest on the creation of credit by the banks. The institution of interest "is neither a necessary nor a sufficient condition for bank's ability to create credit". "A switch over from interest to profit-sharing in banking will be greatly facilitated if it is accompanied by other changes involved in an Islamic transformation of Society". The author concludes that the idea of profit sharing banks has already caught on: a number of reformed banking institutions have recently come up. e.g. the Dubai Islamic Bank is making good progress mainly through credit participation in trade and industry. The two Islamic banks at Cairo and Khartoum are too recent to show any progress. "the case of reform becomes stronger and the one suggested by us increasingly merits serious consideration".

ISLAMIC FUNDAMENTALISM, Economics, Pakistan

72. ENGINEER (Asghar Ali). Islamisation of economy in Pakistan: Some comments. *Link*; 24, 40; 1982, May 16; 27-28.

"President Gen. Ziaul Haq had declared Pakistan to be an Islamic state and in order to convince the people of Pakistan of his intention of Islamisation, he began to enforce the criminal code of Islam as if the essence of Islam lay in enforcing its criminal code". The islamisation process also includes the Islamisation of economy. A committee, of expert economists was set up to look into the economic system of the country and also suggest measures and shortcomings of the Pakistani economy. Beside other things, the concept of riba (interest) is much discussed about these days in the Islamic world. The author is of the view that is entirely wrong to maintain that interest is the source of all evil in an economic system. "The Committee's views (which was appointed by the Government of Pakistan) are quite radical looking to the highly conservative economic views prevalent through out the Islamic world today led by the Saudi monarchy (excepting those countries which have chosen the socialist path). In view of this the views expressed by the committee are quite refreshing and so must be welcomed.

ISLAMIC FUNDAMENTALISM, Economics, Zakat

73. FARIDI (F R). Zakat and fiscal policy. *In* Khurshid Ahmad, *Ed. Studies in Islamic economics*, Leicester, Islamic Foundation; 1980; p. 119-30.

Present paper is an attempt to analyse *Zakat* as the irreducible minimum ingredient of the fiscal policy of an Islamic state. At the very outset it rejects the popular stipulation that fiscal management in an Islamic state is coterminous with Zakat: On the contrary, it assumes the permissibility and also occasional desirability of additional mobilisation of resources for state production of "social good" in an Islamic society. It attempts to incorporate *Zakat* in the over all framework of fiscal policy and studies it as the "leading sector" of the broader complex of revenue expenditure pattern of public authorities. *Zakat* is treated here only as an "economic variable", though its religious importance cannot be minimized, and will be occasionally referred to during the course of this study. The general economic significance of *Zakat* happens to be directional, and normative. It defines the norms of economic activity, also of fiscal activity as a sub-section thereof and determines, through its effects on economic variables, flows and magnitudes, the direction along which the economy is desired to move.

ISLAMIC FUNDAMENTALISM, Economics, Zakat *and* Social Security

74. HASSAN ZAMAN. Islam's system of Social Security: A short Survey. *Voice Islam*; 7, 29; 1983, November 23; 7-11.

Author points out *Zakat* is one of the five pillars of Islam. In Islam wealth is a means to better life and not an end in itself. Islam emphasises on *Sadqa* (Charity) but *Zakat* is duty for every Muslim. In Islam *Zakat* purifies wealth and since it goes spiritual it also increases your wealth. It represents the right to social security for the needy. *Zakat* also combines the concepts of moral and social obligations. A person who does not pay *Zakat* is not a true Muslim. *Zakat* is not a matter of legislation, it is a great co-efficient of production and prosperity. It is the compulsory system of social security in Islam. It is different from the Sadqa and charity. Unfortunately Muslim countries have yet to organize the system of obligatory social security as envisaged in *Zakat*.

ISLAMIC FUNDAMENTALISM, Education

75. ABU AALI (S A). Islamic education: A means towards self-actualization. *In* Muhammad Hamid Al-Afendi and Nabi Ahmad Balach, *Ed. Curriculum and Teacher Education*; Jeddah, King Abdul Aziz University 1980, p. 54-62.

Emphasises that Islamic education, with its outlook towards the universe, towards man who is responsible for inheritance of the earth, and towards the method of acquiring knowledge, which includes research, analysis, addition and evaluation is a means to self actualization of the individual and to realisation of the importance of cooperation in society.

76. ADELEYE (Mikail Clasupo). Islam and education, *Islamic Quart*; 27, 3; 1983, 140-48.

Islam gave so much importance to education that the very first verse of the revelation begins with the vital word *Iqra* meaning read or recite. Islam is said to be a way of life and it teaches how to do all things in life which provide a basis for education and learning. In the early stages of Islam education was confined in mosques but slowly higher institutions of learning were established in Cairo, Baghdad, and Damascus. Through these institutions, a number of Muslim scholars were produced who became well known to the world. Today the centre of learning has moved to formal schools and modern method are used. In addition the students still learn more at home about Islam under the local Islamic teachers. Thus Islam is primarily based on learning and the acquisition of knowledge, which are essential virtues in the Muslims.

77. AL-AFENDI (Muhammad Hamid). Towards Islamic curricula. *In* Muhammad Hamid Al-afendi and Nabi Ahmad Baloch, *Ed.*

Curriculum and Teacher Education; Jeddah, King Abdul Aziz Univ, 1980, P. 3-20.

"Educationists who are found of modern educational terminology and who feel attracted by apparently new educational theories will find that all that is truly Islamic is really modern, unique and genuine. The Holy Quran and the Sunnah contain many references to educational theories and methodologies which should be interwoven into the curricula of Islamic education in Muslim countries. The author also described some of the distinctive features of Islamic educational curricula in this paper.

78. CHOUDHURY (Masudul Alam). Model of educational planning and development in Islamic perspective. *Islamic Order*; 5, 3; 1983; 15-35.

The main objective of this paper is to set out the educational philosophy in Islamic perspective and then to rationalize an educational planning system on the basis of this philosophy. The various aspects of the educational planning system are shown to be the spring board of a new system of inputs to socio-economic development now being considered in its Islamic perspective. From such a rationalization emerges a new dimension in the theory of human capital. This paper initiates an analysis in the direction of this new theory, that could come about in the Islamic frame work of education planning and socio-economic development. The main contribution of this paper is to initiate an approach towards a theory of human capital in Islamic perspective.

79. GHULAM SARWAR. Islamic concept of education. *Radiance*; 19, 45; 1984, Mar, 18-24.

Author is of the opinion that it is an axiomatic truth that Islamic education demands total Islamisation of the whole range of studies. Hence there is a need of re-classification of knowledge and redesignation of curricula. In the words of author "In order to Islamise the prevailing system of education in the Muslim world and to repair the damage done to Islamic culture during alien rule, the need for preparing and training the right type of Muslim teacher whose commitment to Islam, its ideology and values is paramount cannot be over-emphasised. In the present period of crisis and transition, the position of the Muslim teacher is more significant than at any other time".

ISLAMIC FUNDAMENTALISM, Education, Art

80. al-FARUQI (Lois Lamya). Islamization through art: Im-

plications for education. *Islamic Cult*; 56, 1; 1982, Jan; 21-36.

Author feels that "we cannot expect every educational establishment in the Muslim world will provide professional training in music, architecture and the visual arts; but it is not too much to suggest that each Muslim country provide, in one of its institution, a sound programme for professionals in the various arts, a programme based on Islamic aesthetic principles. The training of school children at all levels will have the effect of creating for the future a discerning audience for Islamic works of art".

"Another means for stimulating the creation of new art products which are representative of the characteristics of Islamic art would be to hold International symposia or seminars for artists which would foster and encourage their interest in the principles of truly Islamic art. Exhibits, lectures and panel discussions connected with the seminars should be designed to show the artists as well as the public, how the various arts can be related to *tawhid*." "There could be no more convincing proof to both artist and non-professionals attending that Islam is operative in every aspect of a Muslim's life."

ISLAMIC FUNDAMENTALISM, Education *in relation to* Society

81. AKBAR MUHAMMAD. Islam and national integration through education in Nigeria. *In* John L Esposito, *Ed. Islam and development*; New York, Syrcuse Univ Pr, 1980, p. 181-205.

Discusses the special significance of educational reform in Nigeria. Muslims constitute the largest religious community in Nigeria. Like Malaysia, Nigeria is faced with building a nation in a multiethnic and a multi-religious context. The most important factor in building and reinforcing Nigerian Muslim identity is the traditional Islamic education system. The author indicates, Islam is often "subject to continuous attack as unsuited to modern society" by those who advocate a national western-oriented educational system as means to national integration. However, in recent times Nigeria's Muslim political leadership and western-educated Muslim intellectuals, a minority of Muslim community, have supported a policy of national integration through education. Given the strength and extensiveness of the traditional and modern elites is inevitable. For the former, such an approach threatens the integrity of their community and way of life as well as their own authority.

ISLAMIC FUNDAMENTALISM, Education, Pakistan

82. HALEPOTA (Abdul Wahid). Islamic Educational System. *Islamic Stud*: 20, 3; 1981; 179-99.

Deals with the constitution of the Islamic Federation of Pakistan and education: the past perspective (New ideals with the advent of Islam and the impact of Islam on post-Islamic world culture); the Quranic concept of education; the foundations and basic principles of Islamic society; Islamic educational values: aims and objectives of the Islamic education (building up a pious mind, devoted to the observance of law and Sharia etc.). The article finally deals with "Islamic education through the study of natural and physical sciences". "The study of physical sciences can become 'Ibada and contribute to the same objective as that of religious sciences', which the author attempts to illustrate by discussing the Islamic conception of knowledge; Imam al-Ghazali's view; Shah Wali Allah's views; Muslim's every day activity which becomes *Ibada* (no bifurcation of life into the so-called religious and so-called secular"); "study of physical sciences can create conviction about the existence of the creator", etc.

ISLAMIC FUNDAMENTALISM, EGYPT

83. ALTMAN (Israel). Islamic Movements in Egypt. *Jarusalem Quart*: 10; 1979, Winter; 87-105.

One of the most significant development in Egypt in the 1970s has been the resurgence of Islam as a political and ideological force. It has been shown that Islam is still the most effective form of consensus in Muslim countries. In Nasserist Egypt the weight of Islam has increased considerably since June 1967 and Egypt has experienced increasing manifestations of Islamic sentiment and a growing interest in its ideology. This process accelerated under Sadat. Since the establishment of Muslim Brothers by Hasan al-Banna the fundamentalist movement gained popularity. Since 1972 efforts have been made to introduce Islamic legislation in Egypt. The Islamic group are also active in determining college curriculum, attempting, for example, to convince authorities to oblige all students to attend classes in *Koran* and *Sunnah*. A group known as the *Mahdist group is* also active for its fundamentalist activities. This article is a comprehensive account of the Islamic movements in Egypt.

84. IBRAHIM (Saad Eddin). Anatomy of Egypt's militant Islamic groups: Methodological note and preliminary findings. *Int J Mid East Stud*; 12, 4; 1980; 423-53.

Iran's Islamic revolutions seems to have taken the world by surprise. The western mass media have subsequently been alarming their readers with warning of Islamic "revival" "resurgence" "rumble" and "anger". The author discussed, in this paper, the recent emergence of Islamic militant movement in Egypt and points out its importance. Because Egypt is the centre of the Arab-Muslim-World. In conclusion, the author says that "its success in dealing with the host of global, societal and individual issues discussed in this paper would enhance Islamic militancy. Its failure, especially from within, and without foreign intervention, would set back Islamic militancy. The vision of establishing an Islamic social order has continued to dazzle the imaginations of all Muslims for ages".

ISLAMIC FUDAMENTALISM, EGYPT, HISTORY

85. AJAMI (Fouad). In the Pharaoh's Shadow; Religion and authority in Egypt, *In* James P Piscatori, *Ed*, *Islam in the political process*, Cambridge, Cambridge Univ. Pr; 1983, P. 12-35.

Traces the history of Egypt from the early times to the present, in this paper. He also discusses about the Egyptian Islam. The author throws light on Sayyid Qutb, Muslim Brotherhood, Hasan, al-Banna and other prominent religious personalities. The author says "even if the dominant political order in Egypt were to come unstuck, it is not likely that Islamic fundamentalists would come to dominate the new world". The whole article is the outcome of the above feeling.

ISLAMIC FUNDAMENTALISM, HISTORY

86. ADAMS (Charles J). Background of the contemporary Is-

lamic resurgence. *Islamic Order*: 4, 1; 1982; 54-69.

Author "has tried to show that the appearance of neo-revivalism among Muslims in the twentieth century is part of the dynamics and vitality of the religious community in response to its greatly changed circumstances in our time". "To increasing numbers of Muslims a polity and a social system drawn from the material offered by their own tradition and based firmly on Islamic precepts has seemed to offer the answer to the survival and well being of the community. In some instances the appearance of these vigorous, legally oriented Islamic movements has expressed itself as protest movements against existing regimes; in others they have been co-opted by Islamicizing regimes or have made outright alliances with those in power. In all cases, however, neo-revivalism should be interpreted as a socially and historically conditioned attempt to answer the great questions that have faced the Muslim community in this century. These are questions of identity, of legitimacy, of dignity, and of purpose".

87. ISRAR AHMAD. Islamic renaissance: The real task ahead. *Islamic Order*: 5, 1; 1983; 61-76.

In this article the author says that the present age is predominated by the western philosophical thought and learning. A series of scientific discoveries led to greater control and exploitation of nature, and a wealth of new inventions made Europe an invincible power. Muslim states of the near and Middle East were also subjugated by western powers. "The West's occupation of the Islamic world was two-fold, military and political as well as ideological and cultural. But since it was initially political the reaction against it were political only. Devout Muslims worked to protect their faith and religion. A number of new political, economic, and religious movements emerged. Twentieth century revivalist movements started almost simultaneously in Muslim countries. They were similar in a number of ways. Egypt's *Ikhwanul Muslemoon* has met almost complete disintegration within the country. The Indian sub-continent's Jamaat-e-Islami fared no better. The author 'thinks that the real cause of the failure of these revivalist movements lies in the impatience of their leaders. Their interpretation of Islam affirms all the religious belief but it lacks the inner state of deep faith in God. "Islamic renaissance can never be brought about without first reviving and indeed revitalizing the faith of a large part of the Muslim community". Author suggests that for this" the most essential task to be undertaken is to launch a high-powered academic movement which brings about a real change in the educated elite and intelligentsia of the society. He also suggests various steps that "must be implemented immediately in order to launch the above mentioned academic and Quranic research movement".

88. JANSEN (GH). Militant Islam: Soldiers of Allah advance. *Current Aff Bull*; 56, 1; 1979; 26-31.

Author points out that in the final quarter of the twentieth century Islam became central to every aspect of life., including politics, in countries with Muslim majority. Author describes many reasons for the aliveness and resurgence of Islam. According to the author "the most serious western challenge to all Islamic countries today is on the plane of culture." The author also name the countries where attempts have repeatedly been made for the Islamic revival. They are Libya, Egypt, Turkey, Iran, Pakistan, Malaysia and Indonesia, Sudan, Jordan and Syria. The length of the list shows that 'militant Islam', so much in the news today, is actually nothing new. The author defines the 'militant Islam' in this article and says that their objective is the Islamic state. He also discusses about the Islamic or Muslim countries and their politics of Islamic resurgence.

89. KHWAJA (Jamal). Religious revolution of the 18th century and Islam. *Islam Mod Age*: 14, 1; 1983, Feb; 20-30.

"In this article—the author shall seek to answer 1. How the rise and development of the natural and social sciences in the modern era have gradually transformed the medieval conception of the nature and function of religion in human society, 2. How this change has gradually led to the emergence of religious existentialism, cultural pluralism and 'permissiveness' in western society, and 3. The imperative need that Muslims should grasp the historical logic of the above developments for autonomously applying them to the great Islamic tradition instead of merely revising the Shariah here and there as part of a programme of adjustment to the modern age.

90. SEN (Samar). Islam rises again. *Seminar*; 246, 1980, Feb; 18-22.

Recognizes the "contribution Islam has made to the world's history and civilization". In spite of the various setbacks "the process of contributing Islamic values and knowledge had continued". However, by the end of the 18th century, religio-political power of Islam had crumbled. The partition of India was another important factor. The establishment of Pakistan and Bangladesh has not solved the problems of the Muslims in the sub-continent. The degradation of Muslims continued. A large Muslim population came under western domination. Different sects appeared determined to modernize Islam but in vain. But, since most of the important members of O.P.E.C follow Islam, Muslim all over the world underwent a sea change as a result of the new found bargaining power of the oil producers. In Iran the picture is more perplexing. Since all that Iran stands for today has been obtained under the banner of Islam, the air is heavy with talk of Islamic resurgence, revival and renaissance. In Pakistan the call to Islam has been used primarily to achieve political end. The author raised a question that "these changes in the Muslim world around us may indeed be considered as revolutionary but do they presage an Islamic renaissance or resurgence of Islamic fundamentalism. Apart from the hurdles which these revolutions have still to overcome, a resurgence would once again bring its contribution to art and literature, to science and medicine, engineering and architecture and to navigation and geography, to the emmence benefit of mankind. Thus what we are witnessing may not be an Islamic resurgence but at least an assertion of pride and spirit after centuries of despondency and disillusionment.

91. VOLL (John O). Islamic past and the present resurgence. *Current Hist*. 78, 456; 1980, April; 145-48, 180-81.

In this article, the author points out, "in a variety of contexts ranging from the monarchies of the Arabian Peninsula to the modernist adaptations of Egypt and the more radical experiments of Islamic republics in Libya and Iran, Muslims are working to create societies that are clearly Islamic and at the same time are effective participants in the modern world". "The readiness of Islamic communities to adopt new techniques and technologies is part of the historical record. However, equally clear is their firm resolve to maintain a distinctive Islamic identity in accord with the message of the Quran and the Sunnah. Whenever this resolve appears to be undermined by modern adaptations, Muslim adapt a more fundamentalist style to reaffirm the validity of Islam.

92. VOLL (John O). Renewal and reform in Islamic history: *Tajdid* and *Islah. In* John L Esposito *Ed. Voices of resurgent Islam*, New York, Oxford Univ. Pr. 1983, P. 32-47.

Author says that in the contemporary period, at the beginning of the fifteenth Islamic century, the *tajdid-islah* tradition has a special meaning and relevance in the Islamic vocabulary of resurgence". The global context of the last quarter of the twentieth century is in many ways specially suited for the re-emergence of the renewalist Islamic traditions. The "resurgence of Islam" is at least in some of its aspects, a utilization of tones and symbols that have deep roots with in Islamic tradition. This does not mean that Islamic movements are inherent by "conservative" or "reactionary". On the contrary, it means that a long standing revolutionary or revitalizing tradition with in Islam is being more fully re-activated. Author also describes the importance of *tajdid-islah* in this article.

ISLAMIC FUNDAMENTALISM, INDIA, HARYANA

93. AGGARWAL (Pratap C). Islamic revival in Modern India; The case of the Meos. *Econ and Pol Wkly*; 4, 42, 1969, Oct, 18; 1677, 1679-81.

It was recently before 1947 that Meos of Mewat practicing all Hindu customs except the circumcision and burial of the dead. They were also proud of their Rajput background, though they were Muslims. Since 1947 a revivalist movement started and the initiative was provided by *Tablighi Jamaat* as a result Meos are abandoning Hindu ceremonies. The riots in 1947 convinced Muslims that it is better and secure for them to become full muslims when their villages were attacked by Jats. In four national elections Meos have elected Muslim candidates. With the rapid development in transportation and progress in education, the Meos are more enlightened but religion is most important factor. They try to interact with Meos and Muslims only. They have established many schools and even colleges and together with these secular schools they have Madarsas and many of these Madarsas are more popular then secular schools in Mewat. Even in 1984 we find Mewat like places where pragmatic adaptation to a social setting, where religion, caste, and linguistic barriers are still strong. But this Islamic revival among Meos is accompanied by an enhanced interest in modern education. And it has stimulates them towards modernization to a far greater extent then the urging of the community development organization.

ISLAMIC FUNDAMENTALISM, INDIA, HISTORY

94. KALIMUR RAHMAN. Indian Muslims: The historical perspective. *Radiance*: 17, 11-12; 1981, July, 26-Aug, 2; 13-17.

With the ushering in of the 15th century Hijra at the fag end of the 20th century A.D, or even earlier, due to the world situation, increasing dependance on oil as the primary source of energy, the attempts at Islamic resurgence through Islamic revolutions in countries with predominantly Muslim population, and their geo-physical contiguity on the world map, more so, because of the untold rich quality reservoirs of oil all such countries contain under their surface at comparatively lesser depths, interest in Islam and Muslim has reawakened and increased. India holding the second largest concentration of Muslim population in the world. The author quotes Mrs Indira Gandhi who in a seminar said "We have always recognized Islam as one of our own religions and Islam will continue to grow and flourish in India". Indian Muslim has never felt differently from his Hindu brother, and shared the same emotions, agonies, joys or sorrows. Indian Muslim is as much Indian as any one else.

95. MISRA (Satish C). Indigenisation and Islamization in India. *Secular Democracy*. (Annual Number); 1974; 59-65.

Explain one of the illustrations of the process of social change is provided by the formation of the Indian Muslim community. Basis of this, was not hatred among the Hindus and Muslims but social forces were more important. Afghans were very rigid in the beginning but slowly they had moved from tribal system to a feudal one. As the immigrant communities suffered indigenisation which adapted them to land of their choice convert communities which accepted

Islam generated a reverse process which may be termed Is-lamisation. Here Islam was presented to them as an improved and purer version of Hinduism. The first 300 years of Muslim rule in India were the years of struggle between the two systems of privilege, the local trying to retain it and the introducing trying to secure it. It was a conflict of land and it could be settled only in the reign of Akbar. It will thus be seen that the real differentiation in culture and out look was not essentially between the Hindu and the Muslim, it was between the upper class Muslim and his lower class co-religionists. Thus they can not be studied in isolation from each other.

ISLAMIC FUDAMENTALISM, INDONESIA

96. KHAIDIR ANWAR. Islam in Indonesia today. *Islamic Quart*; 23, 2; 1979; 99-102.

"The Muslim movements, which have made such marked progress in different parts of the Islamic world over recent years, has not left Indonesia unaffected. Muslim make up approximately ninety percent of the country's population, and their determination to revive their religious institutions and give a fresh impetus to Islam through out the archipelago reflect the world wide Islamic revival; reform of the educational system has been among their priorities, and there is evidence of a general desire to cast development programmes in an Islamic mould". This article is the expression of the same feeling discussed above.

ISLAMIC FUNDAMENTALISM *in relation to* IMPERIALISM

97. ALTAF GAUHAR. Islam and the secular thrust of western imperialism. *In* Salem Azzam, *Ed. Islam and contemporary society*, London, Longman, 1982, p. 213-30.

An attempt has been made in this paper to define imperialism, its brief history and then to formulate the issues confronting the Muslim nations. The main target of western secular imperialism is the Islamic faith and culture every where including Europe, the Middle East, South Asia and Africa. The destruction of Muslim cultural identity is the principal aim of western imperialism. The author has tries to identify those western cultural influences which continue to dominate and govern the life of Muslim world and alienating them from their own tradition, making them strangers to their own culture.

ISLAMIC FUNDAMENTALISM in relation of MEDIA

98. ABDUR RAHIM. Muslim world and the western Press. *In* Anwar Moazzam, Ed.. *Islam and contemporary Muslim world*, New Delhi, Light and Life, 1981, P. 181-202.

In fact, most journalists from developed countries familiar with the problems discussed here (in this article) would admit that there are significant shortcomings in the international reportage of the Third World. The attitude of the western Press towards Muslim world or countries are "discriminatory", "biased" and even fabricated. There is no denying the fact that the status and importance of Muslim countries has undergone a sea change. To day, many of the Muslim countries occupy a pre-eminent place in the world economic system because they are abundantly mineral-rich. "What worries the west is the fact that the Muslim countries today dictate the terms of international trade and finance". Western Press has been attracted to cover events of Muslim countries "because their national interests are being directly affected by the development in the Muslim countries". Thus "by and large, the quantum of coverage notwithstanding, the quality and tenor of the news of Islamic resurgence in the Western Press has been basically subjective, their analysis always keeping in the foreground western national interest. Since the beginning of 1978, the Islamic crescent from Morocco to Iran and from Pakistan to Indonesia has been witnessing a great resurgence this revivalism poses a danger to the world. To over come above described situation the author also put forward solution to fighting the mass media against the Muslim countries.

99. MAHATHIR BIN MOHAMMAD. Western Media and Islam. *Islamic Order*, 5, 3; 1983; 94-96.

There are almost fifty independent Muslim states in the world. A gradual improvement in their economic position has aroused great interest in the role of the Muslim world in the creation of a new world order. Unfortunately the attention given to the Muslim world especially by the western media is far too often biased. The efforts of Muslims to make the principles and ideals of their faith play a more meaningful role in the development of a just and prosperous society is portrayed as the work of as strange group of people they refer to as the fundamentalists. Moreover "Most of the so-called western experts on Islam are extremely ignorant of even the most basic of Islamic teachings". In the coverage of the resurgence of Islam, the western media gives unfavourable labels to Muslim. Besides various other labels, labels such as extremist versus moderate and fundamentalist versus progressive, are used with ulterior motives.

100. STODDARD (Philip H). Themes and variations. *In* Philip H Stoddard and others, *Ed. Change and the Muslim world*, New York, Syracuse Univ. Pr, 1981, P. 11-21.

Points out that "the world of Islam, after fourteen centuries of development, has forced itself on the American consciousness only recently. Before the sudden rise in the price of oil in late 1973 and 1974, Islam was largely unknown to most of the west. Since 1974, however, many Americans and other Westerners have found themselves deeply troubled by developments in the Islamic world. Western world press coined the words 'Islamic explosion' 'Islamic whirl wind' of militant Islam, 'Crescent of crisis' 'Islamic resurgence' 'Islamic revival' Islamic fundamentalism etc. for the rise of Islam in the contemporary Muslim world. "The mediated to see Islam as an anti-American monolith and Islam's attitude toward America as the yard-stick by which to measure its validity. Descriptions of trends in the Islamic world generally have ignored the diversity of Muslim peoples and the debate of fundamental issues that permeates the Islamic world".

ISLAMIC FUNDAMENTALISM *in relation to* MEDIA, AMERICA

101. Vonder MEHDEN (Pried R). American perceptions of Islam. *In* John L Esposito, *Ed. Voices of resurgent Islam*, New York, Oxford Univ. Pr.1983, P.18-31.

Author points out that the public's perception of Islam tended to be characterized by ignorance, confusion, and misinformation. The average citizen knows little of the Five Pillars, the expansion of Islam to Southeast Asia and Black Africa, differences among sects and the basic fact that the majority of Muslims live outside the Arab world. Islam and its causes have generally not received a sympathetic hearing and have been the subject of considerable criticism related to such issues as the Iranian crisis, the Black Muslims in the United States or the Arab-Israeli conflict. Author's basic thesis in this study is that views of Americans towards Islam are framed by attitudes toward issues that involve Muslims. These issues are generally not fundamentally religious in content but help to establish stereo-types that are transferred to Islam itself.

ISLAMIC FUNDAMENTALISM *in relation to* MODERNISM

102. ABDUL MOGHNI. Islamic fundamentalism and modernists. *Muslim Wld League J*; 9, 6; 1982, April; 20-25.

Author says that the latest developments in the politics of Iran, Pakistan, Afghanistan, Turkey and Malayasia, have chiefly provoked the apologists of modernism to decry what they call revivalim in the world of Islam in our time. These reactionaries against Islamic revolution in the present day world have coined a curious term 'fundamentalism' to denounce the religious upsurge in the Muslim countries. If called to explain the term, the coiners would say that it means going back to original concept of society propounded and practised by early Muslims in the days of the Prophet and his companions fourteen hundred years ago. As such, the modernist assume that social reconstruction on a purely Islamic line is, as if, time-barred. The author further says that "there are definite indications of a universal renaissance under Islam in modern times". All the western ideologies of present day world have totally failed mankind". The light of Islam has begun, again, glimmering to dispel the growing darkness of modernity, just as it had dispelled the dark of the antiquity and led the world to modernity. The rise of what is being called 'Islamic fundamentalism' must be viewed and reviewed in this perspective".

103. ABDUL MOGNHI. Muslim attitude towards concepts. *Radiance*; 9, 21; 1971, Dec. 5; 5, 12, 14.

In this paper the author says that "the only valid attitude of the Muslims to the modern concepts would be that the postulates of Islam should be applied to these and they betested on the touchstone of Islam. Then purely on merit, whatever is found compatible with the spirit of Islam, be accepted and supported. The author ask a question 'Are the Muslim intellectuals ready to play the universal role of Islam in the modern age and the Indian society. Have they got the necessary courage and capacity for the momentous enterprise? They should not fail to note that they cannot afford this failure now, that their liberation lies in their being able to play the role of Islam, at its fullest and purest in the modern times.

104. BEN BELLA (Ahmed). Future of Islam depends up on Muslim response to modern challenges. *Radiance*; 19, 36; 1984, Jan. 15-21.

Author in this article says that we are living in grave and dangerous times. We are at a moment in history when the world and its systems are crumbling; when the Muslim world is called upon to respond to the challenges; and when under the call of Islam, the world is seeking to break the iron yoke which suppresses its will and seeks to rob it of its dignity. The answers which Islam gives today, and will give tomorrow to these challenges will decide our future.

105. ENGINEER (Asghar Ali). Islam and reformation. *Islam Mod Age*; 9, 1; 1978; 86-95.

Author is concerned with the pressure for change of Islam in the context of modern industrialized society. He points to some important Islamic intellectuals who advocated reforms (Mohd Iqbal, Mohd Abduh, Rashid Rida, Taha Husain) and mentions the opposition of conference (e.g. March 1976 at Mecca). The author himself does not "view Islam as a system which had provided set answers to all our problems long ago, but as a faith whose universal values, as a historical project, are valid even today, and, which with the reformulation of religious truth so as to be meaningful in the modern world, can be realized even today, in fact in times to come too, if the process so continues with intellectual vigour and sincerity of purpose. It is for the Muslim intelligentsia, rather than the Ulama to ponder over and solve the problems confronting Islam and the Muslims of today, simply being proud of one's heritage is not enough; one should feel duty-bond to contribute to it too".

106. FAZLUR RAHMAN. Islam: Legacy and contemporary challenge. *Islamic Stud*; 19, 4; 1980; 235-46.

Discusses two basic factors "whose constant interaction is ideally the source of all Islamic dynamism. a. the moral-spiritual factor, the values of the Quran, denoted in the Quran by the key term *Taqwa* ("conscience") b. "the community that surrenders itself to God (Umma Muslima)". Thus individual conscience and the collective will to act, *taqwa* and the community are twin pillars upon which-according to guidance given in the Quran and Sunna - the edifice of Islam rests. Attention is paid to the fact that the leaders of the Sharia made practically no distinction between ethics and law (conservatism) further to the advent of the Wahabi and the Indian reformist movements of the eighteenth and nineteenth centuries and to "Islamic modernism". "It was the Muslim modernist, not the fundamentalist or the traditionalist, who recovered the integral Islamic legacy of the earliest days" (Islam's very modes of worship having social and political dimensions). Islamic modernism, however, "has not as yet succeeded in the Muslim world, and for the time being at least, appears to be submerged under a storm of what I call neo-fundamentalism, so much so that even its strength appears to be its weakness". The author sees the modernist's basic fault as his proceeding selectively, not treating "the Quran as a whole".

107. NASR (Seyyad Hossein). Reflections on Islam and modern thought. *Islamic Quart*; 23,3; 1979; 119-31.

In conclusion, the author says, "It is necessary to mention that the reductionism which is one of characteristics of modern thought has itself affected Islam in its confrontation with modernism. One of the effects of modernism upon Islam has been to reduce Islam in the minds of many to only one of its dimension, namely the Shariah and to divert it of those intellectual weapons which alone can withstand the assault of modern thought upon the citadel of Islam. The Shariah is of course basic to the Islamic tradition, it is the ground upon which the religion is based. "the successful encounter of Islam with modern thought can only come about when modern thought is fully understood in both of its roots and ramifications and the whole of Islamic tradition brought to bear upon the solution of the enormous problems which modernism poses for Islam".

ISLAMIC FUNDAMENTALISM *in relation* to MODERNISM, CRITIQE

108. AFTAB ALAM. Understanding Islam's role in modern world. *Mainstream*; 19, 10; 1980, Nov; 9-10.

Author says that "Islam never provided, nor is it capable of providing any politico-economic system". It will be a serious error to interpret the problems of Muslim peoples and their responses to them in terms of the religion of Islam. To understand Islam in the modern world, one has to take into account the interaction between Islam and politics, especially in the background of the phenomenon of colonialism. The world of Islam also came under subjection by the European imperialist power. The pattern of further historical development was also similar in other countries, European education leading to national awakening and national movements and freedom straggle. This also led to a situation highly conducive to reaction and religious revivalism. This is also the reason for the emergence of Islam in the third world. In Iran it is the basis of the struggle of the people against a barbarous monarch. In Pakistan it is a tool for legitimating the rule of the junta—In Egypt it·is an effort to promote Ikhwan's reactionary politics vis-a-vis Egypt's national revolution. Thus it is socio-economic and political conditions that decide and determine the meaning of Islam and Islamic fundamentalism and not the other way round.

ISLAMIC FUNDAMENTALISM *in relation to* MODERNISM, INDONESIA

109. TAMNEY (Joseph). Modernization and religious purification: Islam in Indonesia. *Reling Res*: 22, 2; 1980; 207-20.

Modernization or more specifically education and urbanization has been associated with secularization. Some writers, however, suggest that modern people are not so much secular but religiously different. In this paper the author test the idea that modernization is associated with the purification of religious life styles, The education and community size are related to the decline of folk religion and to a net increase in the proportion of Muslims who are active religious purists.

110. ISLAMIC FUNDAMENTALISM, *in relation to* MODERNISM, IRAN

ANTES (Peter). Iran: An Islamic example for religions in the remaking. *J Dharma*; 5, 4; 1980; 372-79.

Summarizes and concludes account of present day Iran by saying that religion has been used to give concrete expression for rejecting the westernized life-style introduced by Shah. "The aspirations of the masses, however, are for the golden age of Islamic welfare such as existed under Muhammad and Ali, while the leading Ayatollahs supported by the Mullahs and their followers proclaims that the solution for all the problems is to copy what they think the historical prototype of an Islamic state must have been. Re-Islamization therefore -as in most of the Islamic countries - is without any innovative theological impact. It does not take into consideration that in any case, our modern world - be it westernized or not - is remarkably different from that of Muhammad and Ali. New aspirations have been developed such as those of emancipation, trade unions, human rights and so forth. If religion in its remaking does not take care of all these, it is certainly going to fail because its fundamental message is to fulfil man's longing for an answer to his vital questions. Admittedly, Iran is a problematic example for religions in the remaking.

111. GRIFFITH (WE). Islam on the March. *Reader's Digest*: 1979, June; 55-59.

In this article the author says that Shah of Iran was forced to flee and Knomeini who was in exile for 15 years returned victorious. The author further says that Knomeini and his followers were armed with something even more formidable than Shah's sophisticated military hardware: A religious ideal. Moreover Iran itself was being caught up in the militant revival of Islamic fundamentalism. Currently sweeping through the world's 600 million Muslims. According to author "Although circumstances differ from country to country, the Islamic revival in general is a reaction to attempts to modernize Muslim nations along western lines. The Shah of Iran will go down in history as the most famous casualty of the Islamic revival. The Iranian middle class was attracted to the Islamic revival as protest against the flagrant corruption, economic inequalities and tortures of the Shah's regime. Actually, Islam is much more than a theology, it is also an idea of how society should be organized. Islam's rising power threatens to shake the foundation of Muslim countries other than Iran. For example, Egypt is under growing pressure from Muslim fundamentalist to turn Egypt into an Islamic state. Militant fundamentalism is a threat as well in oil rich, pro-western Indonesia and in Malaysia, the world's leading tin and rubber producer. So is the case with Syria and Iraq. In Pakistan, the fundamentalists emerge as a major loser to Islamic ferment. Fundamentalists are strongly anti-Soviet because of the atheistic nature of Communism. Khomeini and other fundamentalist insist that they can promote industrialization while maintaining a traditional Islamic society. The author raised a question that "what should the United States do in the face of the Islamic revival ?" "Adjust to it as best as we can" or "adopt stringent measures to conserve energy and lessen our dependence on oil from Islamic countries.

ISLAMIC FUNDAMENTALISM *in relation to* MODERNISM, SENEGAL

112. CREEVEY (Lucy E). Religion and modernization in Senegal. *In* John Esposito, *Ed. Islam and development*; New York, Syracuse Univ. Pr, 1980, P. 207-21.

The author tests the validity of the hypothesis that as Senegal modernizes politically, economically, and socially religion and religious attitude should modernize. The author analyses the interaction of Islam and change in the 1970s, during which time two pattern seem to have developed indicating the modernization in Senegal has had different results. On one hand, it may not be rendering society more secular but rather spreading a regularization of Islam. On the other Senegalese Islam's heavy reliance on superstition and irregular Islamic praxis seems to be disappearing and giving way to a stricter observance of Islamic belief and practice. The author concludes that the long-term effect may well be the differentiation of religious and secular roles so that *marabouts* will come to be regarded primarily as religious teachers.

ISLAMIC FUNDAMENTALISM *in relation to* PAN-ISLAMISM

113. ENGINEER (Asghar Ali). Arabian and Pan-Islamism. *Man*

Develop; 5, 3; 1983, Sept. 99-108.

In its essence and origin, Arab nationalism is a feeling and the Arab nation and its ties to the Arab homeland are the result of that feeling. It is the common bond of sentiment which has enabled the Arabs to lay claim to a single nationalism which pays no regard to the artificial division of the Arab home-land. The feeling that is Arab nationalism is the result of a combination of factors, such as economy, geography, history, language and religion, which have drawn the Arab together. Even Arab unity which is comparatively more realistic than Islamic unity, is still only a dream. The Arabs are hopelessly divided.

114. JAIN (M S). Muslim fundamentalism and Pan-Islamism. *Manthan*; 4, 4; 1983, Feb; 51-56.

Author says that the "two Islamic movements have become, during the last two decades, significant on account of their international bearings. The first is Pan-Islamism and the second is Islamic fundamentalism". In the opinion of the author "both have sought inspiration from certain values enshrined in the Quran and have been put forward by traditional and conservative elements". While concluding this paper the author says "Thus Muslim fundamentalism, Pan-Islamism and Islamic Da'wah have been complementary aspects of a single urge to preserve the political interest of the conservative section of the Muslim society. All these movements operate to assert the 'Islamic personality in political affairs at national and international levels. The basic under currents and main springs of these movements are political. These forums have been forges for waging a political battle for the preservation of a position of leadership for the conservative interests under the garb of preservation of Islamic corporate personality".

ISLAMIC FUNDAMENTALISM *in relation to* POLITICS

115. BROHI (Allaahbukhsh K). Islam, its political and legal principle. *In* Salem Azzam *Ed. Islam and contemporary Society,* London, Longman, 1982, P. 62-100.

In the present paper an attempt as been made to present the essentials of theory and practice of politics in Islam with a view to enabling the reader to appreciate the peculiarities of the lines of thought along which Muslims have solved the problems of politics and the extent to which they, in principle, seem to have contributed significantly to the development of those socio-economic institutions which we associate with and are sanctioned by their religion.

116. ENGINEER (Asghar Ali). Islamic fundamentalism and political motives. *Mainstream*; 21, 14; 1982, Dec; 4; 13-17.

"In the world of politics not wards and terms but intentions are very important. One must not be misled by the terms politicians use but try to probe the intention behind those terms. Islamic fundamentalism, Islamic State, Islamisation etc. are some of the terms which have been freely used by the ruling politicians in a number of Muslim countries since the late seventies. Also, the western media were no less enthusiastic, although for their own reasons, to popularise these terms. The author has tried to "thoroughly examine as well as critically evaluate the real socio-political implications of what has, by and large, come to be known as Islamic fundamentalism, with special reference to Pakistan. Before us there are three countries most vehemently talking about inforcement of the fundamentals of Islam, namely Saudi Arabia, Iran and Pakistan. It is not difficult to see that Islamic fundamentalism is one more powerful and deadly weapon in the armoury of politicians to perpetuate the status quo, with different nuances depending on the situation. Similarly "Islamic fundamentalism", a term coined by the western media, has also certain deep-rooted implications and provides a perceptual frame-work with in which Arabs and Islam are sought to be fitted. "Islamic fundamentalism today is being used as an umbrella term by the vested interest and politicians in Islamic countries. Its real nature, however, would depend on who uses it and for what end. In other words, one must thoroughly examine the politics of Islamic fundamentalism in a particular country before accepting or rejecting the claims made under its cover.

117. HUDSON (Michael C). Islam and political development. *In* John L Esposito, *Ed. Islam and development*, New York, Syracuse Univ. Pr. 1980, P. 1-24.

The author notes in this article that Islamic resurgence brings into sharper focus a fundamental disagreement between the apparent desires of many Muslims (for whom political development is unthinkable without Islam) and the conventional wisdom of western social science (which has maintained that Islam is at best an impediment to political development). Author sees a broad spectrum of roles for Islam ranging from more fundamentalist religious states like Iran to that of Modern Egypt. A common assumption in development theory has been that modernization weakens religious traditions. The reality in many Muslim countries contradict this facile presupposition. The author correctly concludes that the Islamic resurgence is not so much the product of mass alienation of rejection of modernization as the re-emergence of Islam as an important component of political ideology. Islam is an instrument espoused both by incumbent governments and opposition forces as they respond to the political exigencies of their countries and try to obtain legitimation and mass support for their programmes and policies.

118. ISRAELI (Raphael). New wave of Islam. *Ind J*; 34, 3; Summer 1979; 368-90.

Recent developments in Iran, where a rapidly modernizing and seemingly stable nation succumbed so easily to a fundamentalist Islamic upsurge, raise many questions concerning the potential of Islam as a political force in the contemporary world. The new wave of Islam is by no means limited to Iran, or peculiar to the Shi'ite Islam. There have been similar though smaller eruptions in other parts of the Islamic world, and, although there is as yet no conclusive evidence of their interconnection, a trend is clearly discernible. Author has tried to answer those questions which has been raised due to a fundamentalist Islamic resurgence, in this article.

119. MOHAMMED AYOOB. Politics of resurgent Islam. *Australian Out look*; 34, 3; 1980, Dec; 300-307.

This paper has two major objectives. The first is to attempt to analyse, however imperfectly and briefly the use of Islam as a political tool or a political ideology by assorted groups leaders and parties with in the Muslim world for the attainment of certain primarily secular objectives. The second objectives of this paper is to look at the broad question of the Islamic world's response to its encounter with the west. The author is of the view that "this is essential if one is to understand the basic nature of Islamic political revival today, and particularly its impact on a still largely western-dominated international system". The author discusses the politics of Islamic resurgence in details and comes to conclusion that "Muslim resurgence, therefore poses no threat except to those who themselves harbour aggressive and expansionist ideas and subscribe to hegemonic designs which they attempt to justify as essential instruments for the preservation of world order".

120. MOHAMMED AYOOB. Revolutionary thrust of Islamic political tradition. *Third Wld Quart*; 3, 2; 1981; 269-76.

Stressed that from the earliest times, there has been present a revolutionary strand in the Islamic political tradition although, as has been the case with most such ideologies, it has remained the ideology of the opposition and not that of the establishment in the Muslim world. The movements of social and political protest during the first millennium of Islam invariably took on religious character. The beginning of Shia Islam, can infact, be traced to such a protest movement which took on the character of a religious Schism. The author discussed the politics of the Islamic resurgence in Muslim countries and various other modern social theories in the present article.

121. MUZAFFAR HUSAIN. Two faces of Islam: Expansionism and revivalism; *Manthan*; 4, 4; Feb, 1983; 57-66.

Discusses about the expansionism and revivalism in Islam in the present article. He says that the use of oil as a political weapon in the seventh decade of the present century whetted the appetite of the Muslim politicians for starting yet another movement for Islamic expansionism. The success of the Iranian revolution, the climate of terrorism and the immense oil wealth of Arabs nurtured the roots of this revivalist movement and as a result, these forces have intensified their activities in all those countries where Muslims are either in a majority or minority. Today the horizons of no country in the world are free from the clouds of Islamic fundamentalism and expansionism.

122. PIPES (Daniel). This world in political!: The Islamic revival of the seventies. *Orbis*; 24, 1; 1980, Spring; 9-41.

The Islamic revolution in Iran has focused western attention on the place of Islam in current politics: has there really been a revival of Islam? If so, what accounts for it? Political action in the name of Islam has indeed increased during the 1970s through out the Muslim world from Morocco to Indonesia. Among many significant factors behind this, the effects of oil price increases through the decade are stressed in this, study. Most beneficiaries of the OPEC boom are Muslims (for example 11 out of the 13 OPEC members have Muslim head of states); oil money has brought them wealth and power; it has also disrupted their societies. In either case, it has made Muslims more receptive to Islam as a political ideal and as a social bond. Establishing a connection between the oil boom and the Islamic revival is the primary argument of this article.

123. VASILYEV (A). Islam in the present day world. *Link*; 24, 24; 1982; Jan, 26; 77-80

"Islam plays an important role in the ideological struggle and social and political life in the present day world, particularly in West Asia". "This role is sometimes enhanced by various factors, both internal and external, such as the Arab countries" conflict with Israel and the growing economic and political influence of wealthy oil producing states particularly Saudi Arabia. The author says that "more and more people in the Muslim countries are becoming aware of the secret mainsprings behind the "concern" for Islam that is being demonstrated by the imperialists and their stooges".

ISLAMIC FUNDAMENTALISM *in relation to* POLITICS, EGYPT.

124. AYUBI (N N M). Political revival of Islam: The case of Egypt, *Int J Mid East Stud*; 12, 4; 1980, Dec. 481-99.

The recent events in Iran, Saudi Arabia, and Afghanistan, and in Libya and Pakistan, as well as the less widely publicized events in Turkey, Syria, Egypt and the Gulf, have stimulated and renewed people's interest in understanding both the role of religion and the religious revival in Middle east. "However, it is the so-called Islamic revival that has drawn people's attention most in the west, owing in part to political and international considerations". "Islam

plays an important role in Egypt". says author. Further-
more, there can be little doubt that what people call the "Islamic trend" is growing in political significance in Egypt
today". "The most important Islamic movement in Egypt is
the Muslim Brothers (Al Ikhwan al Muslimun). The author
says "Ideologically, there are not many basic differences between older Ikhwan - the fundamentalists and the newer
Islamic organization that we are calling neo-fundamentalists". In every day political life, the presence of the new
Islamic societies is most noticeable. The author concludes
the article with the following that the resurgence of Islam
may not be completely attributed to purely "Islamic"
reasons; and in this Egypt may not be a unique case among
Muslim countries".

125. WARBURG (Gabriel R). Islam and politics in Egypt, 1952-
80. *Mid East Stud*; 18, 2; 1982; 131-57.

The author first considers Egyptian Islam in its attitude to
internal politics, then the role of Islam in contemporary
Egyptian foreign relations, and finally pays much attention
to the populist Islamic movements (the Muslim Brethren
and the Neo-Mahdist movements). As for the latter, it is
concluded that especially since 1973, fundamentalist Islam
of the more militant brand has made headway in all sectors
of Egyptian society. The major reasons for this revival are
to be sought in the decline of other ideologies, such as
Socialism, Marxism or even Pan-Arabism, all of which have
been largely discredited. Moreover, while it is easy for the
regime to denounce the above ideologies as alian to Islam,
it is much harder to suppress those who preach fundamentalism." While Neo-Mahadist movements are "probably too
weak to endanger the regime, their extremism and
popularity may drive the more cautious and better organised Muslim Brethren, to follow a more extremist line.
This in turn, may bring to an end the period of peaceful
co-existence between the regime and the Brethren whose
support is believed to be widespread and effective."

126. YADLIN (Rivka). Rise of political Islam: Contemporary attitudes to Islam in Egypt. *New East*; 29, 1-4; 1980; p. 113-20.

An illuminating feature of the contemporary surge of Islam
is the growing weight of its socio-political component. The
concept of din wa-dawala (religion and government) is

gaining ground outside fundamentalist circles. Support for the establishment of a Muslim government has recently been forthcoming also from the "Muslim Left". The Muslim Brothers and Muslim student groups are more moderate or cautious — in stating their aims. Their efforts are directed towards establishing the Sharia. In the spectrum of attitudes to Islam, one main variety is the religious personal one, focussing on faith and morel behaviour (emphasis on an international Islamic order). This approach enjoys great popularity. Diametrically opposed to the aforementioned trends are the liberals. In fact, secular modernism has invaded the stronghold of the Sharia personal status law. Nevertheless, there can be no doubt that fundamentalist attitude are winning supporters in wider circles. The surge of Islam occurs at a time of acute social and political crises and a renewed search for identity in Egypt of all available solutions, it is the only one that is both comprehensive and authoritative.

ISLAMIC FUNDAMENTALISM *in relation to* POLITICS, INDIA

127. ENGINEER (Asghar Ali). Situation of Indian Muslims today. *Mainstream* (Annual); 1981; 67-71.

In this article the author says that "Muslims constitute an important minority group in the total population of India. The problems of our country can not be understood in their totality without understanding the situation of this important segment of the population." It would be necessary to understand this question on intellectual plane. "On intellectual plane what has disturbed the non-Muslims most is that attitude of Muslims has been described as separatist the routes of separatism should be sought in the situational context, rather than in the religion of the community. Since we are discussing the question of separatism among the Muslims, we should also understand and appreciate the difference between the nature of separatism before and after partition. It was aggressive before partition it is more of withdrawal today from the main arena of struggle. A large section of the Muslim elite, especially from U.P., migrated to Pakistan, leaving a great void. Economically too, with the abolition of Jagirdari system, the Muslims lost their domination. In this grim economic situation where Muslims are extremely backward and mere or below the poverty line, it would be surprising if they were not as tradition bound and conservative as they are today. To them traditional religion is not only more meaningful but is also a powerful source of solace. The politics of religious identity has come to play a very important role among Indian Muslims today. The politics of religious identity has given undue prominance to the conservative Ulema. There is yet another important factor which must be taken note of the fundamentalist movement in the Islamic world. It is not true that Indian Muslims have reacted vary enthusiastically

to the fundamentalist movement. Imtiaz Ahmad is not far wrong in pointing out that "India has remained totally untouched by the recent developments in Iran and Pakistan as well as in other countries of the Islamic world.... the recent resurgence of Islamic revivalism (there) has not struck a sympathetic chord among Muslims in India. The author opines "that politics of religious identity would thus play a more prominant role pushing economic issues to the background".

128. HEIDRICH (Joachim). Islam as a political factor in India; An evaluation. *Link*; 26, 1; 1983, Aug, 15; 85-99.

Those who profess Islam in contemporary India constitute more than eleven percent of the countries' total population. But figure apart, the role of Islam in the South Asian Subcontinent and in India has attracted considerable attention. A study of the political role of Islam would include the question as to how far international trends within the "Muslim world" get reflected in a country with a sizeable Muslim population. I might be asked whether there are boundaries and which features could be distinguished as specific of the Indian scene. It is attempted in this paper to discuss a few aspects of the vide-ranging subject. In the end the author says that "in the recent years ideas of Islamic fundamentalism were sought to be cultivated among Muslims in India through conservative organisations". "Linked with hostility to social progress, the propagation of Muslim fundamentalist ideas in contemporary Indian public life is helping to strengthen internal and external forces of conservatism and reaction". "From the view point of India's experience the attempts to promote communalist ideologies and to foster tendencies of religious revivalism ran counter to the goals of democratic advance and proof to be detrimental to the anti imperialist cause".

129. MOIN SHAKIR. Religion and politics: Role of Islam in modern India. *Econ and Pol Wkly* (Annual Number): 14, 7-8; 1978, Feb; 469-74.

Examines the political and sociological dimensions of Muslim communalism in India. While actual voting behaviour of the Muslim masses brings hope to those sections of the Muslim elite who, however faintly, are striving to secularise Muslim politics, the obstacles in the path of such secularisation are many. The article briefly touched upon these obstacles, as also on the failure of Muslim intellectuals to present a comprehensive analysis of the Muslims situation in India without recourse to myths and mystification.

ISLAMIC FUNDAMENTALISM *in relation to* POLITICS, INDONESIA

130. ROFE(Husein). Islam's role in Indonesia; Insja'llah-God willing. *Far Easter Econ R*; 59, 14; 1968, April, 4; p. 26-29.

The surprising show unity in the people's congress by Indonesia's major Muslim parties in an attempt to restrict President Suharto's power, gave an indication that the country's main religious group is at last showing the ability to present a united front on political matters. Although Suharto is able to count on continued wide spread support among regional and non party delegates, it is becoming obvious that he cannot disregard the demand of the Muslim faction. Prompting the aggressiveness on the part of the Muslims is a growing belief among their leaders that the country's only path to economic stability lies in the application of Islamic principles.

131. SAMSON (AA). Islam in Indonesian politics, *Asian Survey*; 8, 12; 1968, Dec; 1001-17.

Assesses the magnitude of influence of (Islam) religion in Indonesia which is not a secular state. Observes that Indonesia is nation where 90% of the people are Muslims but all attempts by Islamic parties to influence national politics have been frustrated by both Sukarno before and Suharto. The Islamic movement is divided into two groups. Although both the groups believe that only Islam is the solution to every problem, they remain aloof from each other and in fighting goes on between them to gain supremacy.

It is probable that the movement they unite, Islam will re-emerge as an acknowledged political power.

ISLAMIC FUNDAMENTALISM *in relation* to POLITICS, INTERNATIONAL

132. KAUSHIK (Brij Mohan). Islamic Revivalism in disarray. *Strategic Analysis*; 4, 7; 1980, Oct; 313-17.

Author points out that the conflict between Iraq and Iran provides yet another evidence to prove that religion can at best play a very limited role in inter state relations. But it is not just in the context of Iraq-Iran war that we have become sceptical of the revival of Islamic fundamentalism. Author says that "We are at variance with the thesis propounded by several authors, particularly those in the United States of America and few in this country also, that the revival of Islamic fundamentalism is around the corner. The impact of religion on the polity of various Muslim countries is no doubt very much perceptible. But it appears beyond the realm of possibility that 750 million Muslims of the world will succeed in building an Islamic monolith just because they are the followers of Islam. Author gives sufficient reasons for this. However author has not suggested in this article that the revival of Islamic fundamentalism is nearly a figment of some one's imagination. Various Muslim countries are under pressure from within to adopt certain measures from the Islamic tenets for their governance. Author is of the opinion "in theory, no doubt, Islam is the only force that can bring as many as 43 countries together." But "it is not supported by history".

133. KRAUS (Jon). Islamic affinities and international politics in Sub-Saharan Africa. *Current Hist*; 78. 456; 1980, April; 154-58, 182-84.

Author points out that "recent events in the Middle East, especially the militant anti-western Islam of Iran's Ayatollah Khomeni, have apparently not stirred the resurgence of Muslim social forces in sub-Saharan Africa.

However, a few African states like the Sudan have discovered that the Arab states try to compell or induce domestic political environments more conducive to Islamic political forces and beliefs." "In sub-Saharan countries whose population is largely Muslims, local and national politics is often influenced or shaped by Muslim social forces and, less frequently, by Islamic faith or law. However, many factors sharply reduce the impact of specifically Muslim groups or the Islamic religion on local or national politics. Where Islamic political and social institutions were well established prior to colonial rule, they often retain much of their power and even expendable."

134. MALIK (Hafeez). Islamic theory of International relations. *J South Asian Mid East Stud*; 2, 3; 1979, Spring; 84-92.

"The development of the Islamic theory of international relations is an integral part of the movement for the modernist reconstruction of Islamic thought in the Muslim world", says author. In this paper "an attempt has been made to delineate the sharp distinction between the Islamic rule and the Islamic faith. The Quran in its pronouncement, and the Prophet Mohammad in his diplomatic relations, clearly indicated that no one is to be forced to accept Islam. Yet all Muslims were enjoined to establish the rule of Islam, because the Prophet maintain that Islam was a just polity."

135. SEALE (Patrick). Islamic revivalists hold secret summit. *Times of India*; 1981, Dec, 23; 6-8.

"Representatives of Islamic opposition movements from a score of countries flew into London last weak-end for a secret three days meeting, thought to be the first held on such a scale." The meeting marks a development of the revival sweeping the Islamic world. The representatives came to give an organisational framework and a common strategy to the largely spontaneous and far-flung movement of Islamic activism.

ISLAMIC FUNDAMENTALISM in relation to POLITICS, MALAYSIA

136. ABU BAKAR (Mohamad). Islamic revivalism and the political process in Malaysia. *Asian Survey*; 21, 10, 1981; 1040-59.

Malaysia is among several Muslim countries affected by the current wave of Islamic resurgence. Revivalist organizations that sprang up some years ago now have nation wide appeal among the Malays. Their impact is felt throughout the country particularly at the Socio-political level. "It is highly probable that the trend towards Islamic orthodoxy could steer this multi-racial country on to a new political course. Several factor account for this upsurge in religious feelings; Firstly it results from a re-education programme amongst Muslim Malays. Secondly the "back to the Koran" movement aided this Islamization process. Thirdly the secularisation of Malays society can be held responsible for this new interest in Islam as it provokes strong reactions from Islamic circles loosing ground. As a result of this "the influence of Islam in Malaysia has become more pervasive then ever before". At the same time the forces acting against it may try to counter act this increasing influence. These forces are not only represented by the non-Muslims in the country but even more by the Kampung folk, many of these still being imbued with the old conception of religion. For the future an ongoing clash of interests can be expected.

ISLAMIC FUNDAMENTALISM *in relation to* POLITICS, MIDDLE EAST

137. CAMPBELL (WR) and DARVICH (D), Global implications of the Islamic revolution for the status quo in the Persian Gulf. *J South Asian Mid East Stud*; 5, 1; 1981; 31-51

    Author observed that in Khomeni's view, none of Iran's neighbors has a truly Islamic government, but some have incorporated elements of Islam. Next to Iran, Algeria may be truest to Islam but still not fully incorporated as agent of the oppressors are the Arabian peninsula states. The Muslim countries are facing a lot of problem in every field, every walk of life. The general remedy for these problems is to implement Islamic regulations just as they were at the beginning of Islam. Author feels that the great powers intention is to break Iran into pieces. He also discuss the global implications of the Islamic revolution for the status quo in the Persian Gulf.

138. HADDAD (Yvonne Y). Arab-Israeli wars, Nasserism and the affirmation of Islamic identity. *In* John L Esposito, *Ed. Islam and development*, New York, Syracuse Univ. Pr. 1980, P. 107-21.

    Author, in the present article, examines the ways in which the wars and Nasser's response during this period to the political realities of Egypt and Arab world contributed to a reaffirmation of Islamic identity in Arab politics. As author observes, the existence and prosperity of the state of Israel is a direct challenge to the Arab Islamic view of the Historical process. The loss of Jerusalem in 1967 intensified the religious significance of the conflict. For most Muslims Israel stands as a symbol of Western Power and a reminder of Muslim impotence. Nassar was able to shrewdly seize upon these concerns and use Islam in legitimating his rule and policies within Egypt as well as establishing his position as a popular Arab nationalist leader abroad.

139. HUMPHREYS (R Stephen). Islam and political values in Saudi Arabia, Egypt and Syria. *Mid East J*; 30; 1979; 1-19.

Observes that Islam is currently playing a striking role in the political life of the Arab world. It is an appropriate movement to consider the way in which Islam may affect the political values and policy choices of the ruling elites in preponderantly Muslim countries, says author. The paper also defines fundamentalism and discussed it in a great length. Author observes that "Saudi Arabia is clearly a critical test case for fundamentalism." "Egyptian and Syrian society have developed cultural ideals and aspiration during the past century which simply cannot be met within the regid frame work of fundamentalism. For Egypt and Syria, whatever obeisance may be paid to fundamentalist norms from time time, the future must be lie with secularism or modernism."

140. SALEM (Elie Adib). Political trends in the Arab world. *Australian Outlook*; 34, 3; 1980, Dec; 308-14.

This is the assumption of the author "that the major trends in the Arab world are largely a function of international politics." The author has examined the future trends in the Arab world in the light of the above assumption. The author discusses the state of affairs in Muslim countries like Iran, Iraq, Sudan, Somalia, Jordan, Yemen and various other Islamic countries of Asia and Africa. The author says that mosque episode in Mecca is a manifestation of Islamic fundamentalism. He also discusses Palestine problem.

141. TIBI (Bassam). Renewed role of Islam in the political and social development of the Middle East. *Mid East J*; 37, 1;

1983, Winter; 3-13.

"The Islamic resurgence, of which the Iranian Revolution remains a dramatic example, indicates a renewed role for Islam in the political realm. This notion implies that while Islam had become politically less relevant before the process of Islamic resurgence that we are now witnessing, Islam, as a prevailing normative system, has never been questioned. In drawing attention to the renewed role of Islam, we merely refer to the re-politicization of the sacred, to the reviving of Islam as a political ideology."

ISLAMIC FUNDAMENTALISM *in relation to* POLITICS, MIDDLE EAST, SOUDI ARABIA

142. PISCATORI (James P). Roles of Islam in Soudi Arabia's political development. *In* John L Esposito, *Ed. Islam and development.* New York, Syracuse Univ. Pr. 1980, P. 123-38.

Describes how "Islam is an abiding, if not also central, reality in Saudi Arabia." Islam has given life and form to the Saudi state from its mid-eighteenth century origins to the present. In the Quran and the Sharia the Saudi family has found legitimation for their monarchy. The author discusses some changes and more importantly, analyses those factors which have enabled this Muslim Society to use Islam to facilitate reforms. Unlike most states in the Islamic world where civil courts and laws predominate the Shariah is restricted to family law, in Saudi Arabia Shariah courts exercise jurisdiction in all spheres of life (civil, criminal and family). More over Saudi law incorporates both traditional Shariah regulations including the Quran-prescribed penalties (hudud) for theft, drinking and adultery, and royally decreed modern laws governing such areas as mining, commerce and social insurance.

ISLAMIC FUNDAMENTALISM *in relation to* POLITICS, PAKISTAN

143. ESPOSITO (John L). Islamization: Religion and politics in Pakistan. *Muslim wld*; 72, 3-4; 1982, July-Oct; 197-223.

Examines "the role of Islam in contemporary Pakistani politics (which) has been described as the process of Islamization, i.e. creating a Islamic system of government (Nizam-i-Islam). This study reviews and analyses the process of Islamization in Pakistani-its laws, regulations, institutions, programmes and policies. The issues and problems that emerge from this process not only contribute to our understanding of contemporary of Pakistan in particular but also address underlying issues and problems faced by other Muslim countries as they seek to Islamize their societies."

144. SHAH (Mowahid H). Pakistan, Islam and the politics of Muslim unrest. *In* Philip H Stoddard and others, *Ed. Change and the Muslim world.* New York, Syracuse Univ. Pr. 1981, P. 59-64.

According to author "Pakistan is of important strategic concern to the West." "It is unique among modern states in that it is the only state to have been founded on the basis of Muslim nationhood." Author puts a question that is the current Islamization in Pakistan, part of the world wide Islamic fervor? and proceeds to answer it. He says "a closer examination indicates the reasons have more to do with internal politics. In particular, Pakistan reflects a heightened official recognition of the effectiveness of Islam as an instrument to retain governmental control." "In fact, Muslim unrest in Pakistan and elsewhere has surfaced for reasons more political than religious. The author has also discussed Zia's objectives and U.S. policies and interest in the context of world wide Islamic resurgence. Author concludes this articles with the following lines "it has been said, with substantial justification, that Islam is not adequately understood in the West. But as before, Islam will be judged in the fourth quarter of the twentieth century more by the conduct of its followers than by its human precepts. The challenge Muslim face, therefore, is to realize the Holy Prophet's revolutionary message by identifying Islam as an ally of progress instead of its foe. In that realization could be beginnings of a true Islamic renaissance."

145. TYLOR (David). Politics of Islam and Islamization in Pakistan. *In* James P Piscatori, *Ed. Islam in the political process.* Cambridge, Cambridge Univ. Pr., 1983, P. 181-98.

Points out that General Ziaul Haq has succeeded to implement the measures of the Islamization in Pakistan. The programme of Islamization which it initiated in Pakistan in 1978 was a direct, necessary and much-delayed outcome of the country's foundation in 1947 as a homeland for the Muslims of the Indian sub-continent. The opportunity that had been missed in the early years of independence had been recreated at a time when Muslim world-wide were asserting the validity of the original formulas of Islam. The author is of the view "that the development before and after the creation of Pakistan in 1947 can only be understood in terms of the extremely complex interrelationship in South Asia between Islam and political power." Besides other developments the author also describes developments in Pakistan since 1977.

ISLAMIC FUNDAMENTALISM *in relation to* POLITICS, SOUTH EAST ASIA

146. KROEF (Justus M). Political texture of Islam in South East Asia. *J Dharma*; 7, 1; 1982; 56-72.

Probably the major determinant of Islam's place in the political life of South East Asian countries is likely to be extent to which the religion is perceived as facilitating economic growth and political stabilization. An upsurge of fundamentalists fervor, however satisfying as a "revenge" tactic towards repressed social strata seeking to combat a ruling elite, is unlikely to last as a meaningful programme of action in a modernizing state. For an indefinite time Islam's theocratic dimension in South East Asia will have its impact as a national security of jurisprudential problem. But it is not a wave of the future. It is stressed that not even in Indonesia or Malaysia is Islamic orthodoxy numerically sufficiently dominant to the extent that it can hope to determine national political life. To be sure, in both countries Islam has a broad symbolic significance, serving as a major cultural reference point. But such factors as the *abangan* orientation (in Indonesia, meaning less consistently observant of ritual, tending to mix Islamic precepts with indigenous, pre Islamic values, being more tolerant), racial differences - particularly the prominance of the Chinese in Malaysia, and, in both countries, a steadily advancing secularization of life styles, will continue to limit greatly Islam's cultural and political roles.

147. TROCKI (Carl A). Islam: Threat to ASEAN regional unity? *Current Hist*: 78, 456; 1980, April; 149-53, 181-82.

76

All the states of the Association of South East Asian Nations
(ASEAN) are deeply concerned with Islam. The history of
Islam in South East Asia is distinct from the history of Islam
in other parts of the world. The region among the last areas
of the world reached by propagators of the Islamic faith.
"The Islamic political movement in Indonesia is possibly
the more matured Islamic movement in South East Asia."
"The status of Islam in Malaysia will undoubtedly be influenced by events in Indonesia." In this paper, the author
points cut that "currently, all the ASEAN governments
continue to hold the balance of power in their respective
states. In each case, however, Islam presented a new area
of increasing tension, both internally and externally." This
paper also discusses the Islamic revival in Malaysia.

ISLAMIC FUNDAMENTALISM in relation to POLITICS, SUDAN

148. CUDSI (Alexandar S). Islam and politics in the Sudan. *In*
James P. Piscatori, *Ed. Islam in the political process.*
Cambridge, Cambridge Univ. Pr. 1983, P. 36-55.

In the summer of 1980, President Ja'far Numayri of the
Sudan published a book on Islam. It is basically an exposition of his views on the social and political role of Islam
and is intended to explain his adoption of pro-Islamic
policies in recent years. The purpose of this paper is to trace
the role of Islam in Sudanese politics, particularly since
Numayri came to power and to identify the evolution of
a consensus calling for the establishment of a Islamic
republic.

ISLAMIC FUNDAMENTALISM in relation to PROSELYTIZATION, INDIA

149. BHATNAGAR (Brij Bhushan). Muslim reaction to mass
conversion. *Manthan*; 4, 4; 1983, Feb; 93-100.

Describes the reaction of the Muslim community of India
regarding the mass conversion at Meenakshipuram. The
author says it is difficult to find any influential Muslim
voice condemning these conversions. Most of the Muslim
politicians or intellectuals either choose to keep mum on
this issue, or the chosenfew, who spoke out, condemned
not the mass conversion to Islam but the Hindu reaction

to it. The author also describes the role played by Muslim communal organizations like Jamat-e-Islami, Indian Muslim League etc., which are striving to achieve the common aim of establishing an Islamic state in India.

150. WRIGHT (Theodore P). Movement to convert Harijans to Islam in South Asia. *Muslim Wld*; 72, 3-4; 1982, July-Oct; 239-45.

Author says "According to some report upto three quarters of the inhabitants (of Meenakshipuram) converted to Islam in a mass ceremony" in 1981. The author has tried to trace the motives and methods for the conversion in this paper. Besides other factors, history of police harassments and humiliating social discrimination by cast Hindus towards Harijans, were the basic factor for conversions. "Report from the Hindu side charge that many kinds of inducement were offered to the villagers including promises of jobs in Arabian Gulf Area." "As with the Moradabad communal riots the year before, there are rumors of Arab oil money providing most of the financial resources needed for these practices." "More credible is the explanation which attributes the initiative for conversions to educated elements within the untouchable. The author also tries to discuss "A question that deserves attention is why this time the choice for Islam rather than for Buddhism." "In a more constructive vein, Hindu organizations, both local and national mobilized to re-convert the lost sheep, or at least to halt the escape of Harijans from the Hindu fold." And "legitimate conversion was very narrowly interpreted by Hindu organizations." "Whatever the intent or hopes of the Jamaat-e-Islami and the converts themselves and the opposite fears of Hindu militant, the small wave of conversion seems to have died down as precipitously as it arose."

ISLAMIC FUNDAMENTALISM *in relation to* RENAISSANCE, EAST

151. ABDUL MOGHANI. Islamic renaissance in modern times. *Radiance*; 16, 39; 1981, Feb., 8; 3, 8, 11.

These days, there is much ado about the resurgence of Muslim nations and the rise of Islamic fundamentalism therewith. But very few people take the care to ponder the reality of the situation, as it is. A lot of pre-conceived nations hamper the understanding of the phenomenon called

Islamic renaissance, in modern times. However, the phenomena can be neither wished away nor ignored. It is now a fact of life these days and has, as such, to be correctly understood and fully appreciated. The Islamic renaissance is, in fact, a part of the resurgence of the East. Fundamentalist leaders emerged from among the people and challenges their modernist counterparts. The people supported the fundamentalist though the modernist had all the material power at their disposal in view of its roots and bearings, the Islamic renaissance may well be called universal renaissance of mankind under Islam, in modern times. The fundamentalists are just radical revolutionaries who are seeking to bring again into vogue the fundamental principal of Islam, in all walks of life. It is certainly a sort of revivalism but there is nothing wrong or strange about it. In the end the author makes it clear that the revival of Islam does not mean the physical re-production of an age gone by. It means only the revival of the fundamental tenets, principles, and values of Islam, keeping intact, even promoting, all the scientific, industrial and technological developments of the present age under the comprehensive and cosmic vision of Islam.

ISLAMIC FUNDAMENTALISM *in relation to* SOCIAL SCIENCES

152. al-FARUQI (Raji). Islamizing the Social Sciences. *Stud. Islam*; 16, 2; 1979; 108-121.

Author firstly points to the rise of social sciences in the West and natural sciences as its prototype. As main shortcomings of Western methodology are seen the denial of reference to a priori data and a false sense of objectivity, A further argument against Western social science is that it "violates a crucial requirement of Islamic methodology" (the principle of the unity of truth)." Study of society cannot be free from judgment and valuation and is therefore subject to the same rigour, or absence of it, as philosophy, theology, law literature and the arts." "Islamization of the social science must endeavour to show the relations of the reality studied to that aspect or part of the devine pattern pertinent to it." Islamic social science can "humanize the discipline and reinstate the humanists ideal in the life of man, the being whom Western social sciences has taught to regard as helpless puppet of blind forces." Finally some recommendations for action are given with respect to (a) human resources (e.g. formation of an association of Islam —committed social scientists), (b) materials of study and tools of research (c) creative works.

ISLAMIC FUNDAMENTALISM *in relation to* SOCIETY

153. AHMAD ALI. Revival of Islamic values in life and literature. *Islamic Order*; 4, 3; 1982; 41-71.

In this article, accepting the challenge of contemporary aesthetic conflicts, the writer has called attention to restore the image of Islam in certain important aspect. The author feels that the real task before us is to meet "the challenges by adopting the scientific world view to the need of the society, in fact prepare the society for adjusting itself to the modern world view." "By way of conclusion may it be said that there are visible signs of awakening in Islamic world." "Any Jihad for the world of Islam today means removal of its short comings, weakness, lack of strength and preparedness, both in the inner and outer spheres." In fact to prepare society for adjusting itself to the modern world view is Jihad. And unless we can re-instate the values of Islam, all our efforts at survival as a distinct entity can only fail or lead to further degradation.

154. CHAPRA (M Umar). Islamic welfare state and its role in the economy. *In* Khurshid Ahmad and Zafar Ishaq Ansari, *Ed. Islamic Perspectives*. Leicester, Islamic Foundation, 1979, P. 195-221.

Islam has set of goals and values encompassing all aspects of human life including social, economic and political. Since all aspects of life are interdependent and the Islamic way of life is a consistent whole, its goals and values in one field determine the goals and values in the other fields as well. This paper seeks to examine the inter-relationship between the economic and political content of the Islamic way of life and discusses the functions and nature of Islamic state in the light of its basic imperatives within the frame work of financial constraints.

155. al-MAHDI (Sadiq). Islam-Society and change. *In* John L Esposito, *Ed. Voices of resurgent Islam*, New York, Oxford Univ. Pr. 1983, P. 230-40.

A critical issue of the twentieth-century Islam has been the question of change and adoption. If the classical doctrine is that Islam is a total way of life, revealed by God and enshrined in Islamic law, than to what extent is change possible? This question is especially poignant today as Muslim revivalists seek to establish more Islamically-oriented states and societies. The author responds to this question in the present article. He analyses the causes of false views of Islam and attempts to demonstrate how Islam foster and accommodates change. While more conservative fundamentalist Muslim look to the past for their paradigms, author argues that there is no particular Islamic political or economic system. Thus, the task of contemporary Islam is to undertake a fresh interpretation of Islam.

156. SAIYED (A.R). Sociological exploration. *Seminar*; 290; 1983, Oct; 19-24.

The present article has sought to examine the "why of the current resurgence of Islam. The answer was provided by focussing attention on the dynamics of rapid modernization, group interests and the linkage of these two with fundamentalism in modern societies. The author reasonably surmised that a total distinction of Islam and modernization will not occur. Excesses of Islamic traditionalism, like

those of modernization, will have to cease some where, some time, and a new equation between the two will have to evolve. Islamic fundamentalism will have to make peace with modernization, particularly with its developmental component. For, irrespective of its force and power, fundamentalism obviously can be generate an Islamic technology or Islamic medicine etc. In this new equation will ultimately depend on the existential conditions in which the Islamic countries find themselves in the near future.

ISLAMIC FUNDAMENTALISM *in relation to* SOCIETY, ARAB

157. LAYISH (Aharon) and SHMUELI (Avshalom). Custom and Sharia in the Bedouin family according to legal documents from the Judaean desert. *Bull School oriental African Stud*; 42, 1; 1979;29-45.

Attempts to disclose interaction between custom and Sharia and to illuminate some of the mechanisms tending to complete the Islamization of a tribal society in process of sedentarization. Custom is increasingly superseded by the Sharia on its own home ground-arbitration. The encounter with the Sharia is namely of historical importance; in prolongs interaction between them, custom contributed many legel elements to Islamic law. In modern times, the Sharia has the upper hand in the encounter. It displays considerable assimilative power; Islamic legal institutions and institutions originating in custom but having acquired and Islamic connotation are gaining more and more ground in customary jurisdiction. In this transition period of Bedouin sedentarization, a kind of peaceful coexistence is preserved between the two normative systems; the dividing line is some times blurred. There can be no doubt that ultimately the Sharia will prevail and custom will recede. The dependence of the Bedauin on the Sharia court is increasing; the arbitrator's ability to complete with the institutionalized qazi is lessening because the latter is aided by state imposed sanctions. It is not only the increased ascendancy of orthodox Islam as a philosophy but also, and mainly, the secular civic exigencies of a modern state, such as licensing and welfare, that cause Bedouin to resort to the Sharia court.

ISLAMIC FUNDAMENTALISM *in relation to* SOCIETY, EGYPT

158. WATERBURY (John). Egypt-Islam and Social change. *In* Philip H Stoddard and others, *Ed. Change and the Muslim World.* New York, Syrocus Univ. Pr. 1981, P. 49-58.

"Since the early nineteenth century, Islam, as doctrine and social institution, has been challenged by foreign political dominance." "Because of its profound importance to Egyptians, Islam has been at once a cause of dissension and a source of unity. The call of fundamentalist Muslims has resonance because it invites participation. At a time many Egyptians feel confused and find themselves facing an uncertain future, Islamic fundamentalism connects them to a tradition that reduced bewilderment." "The important and the power of Islamic fundamentalism lie in its ability to destabilize a regime and to help bring it down by denying it the religious mentle that remains and important source of political power." This article conveys the same feeling and discusses the resurgence of Islam in Egypt in detail.

ISLAMIC FUNDAMENTALISM *in relation to* SOCIETY, IRAN

159. ABIDI (AHH). Iranian revolution: Its origin and dimensions *Int Stud*; 18; 1979; 129-61.

"The revolution in Iran encompasses distinct human, social, religious, economic, political, legal and constitutional issues which, though not finally resolved, are not peculiar new phenomenon but the latest phase of a process of politico-constitutional evolution which begun during the second half of the nineteenth century." This article is the expression of the above feeling where the author in the conclusion says "Iran has just passed through the most crucial and ex-cruciating phase of its revolution." "Which saw the collapse of the monarch and the monarchy and the establishment of Islamic republic. The next phase in such a revolution is normally much more complex as it involves the intricate task of pacification and consolidation."

160. KEDDIE (Nikki R). Iran: Change in Islam; Islam and change. *Int J Mid East Stud*; 11, 4; 1980; 527-42.

Author gives special attention to the recent history of Iranian Shi'i Islam in an attempt to overcome the relative scholarly neglect of majoritarian thinkers among the Ulema and the over emphasis on Western-influenced reformers. "That neglect has characterized even studies of Iran, although less drastically than it has that of Sunni countries owing to the obvious importance of the Iranian Ulema in nineteenth and twentieth century Iranian history". He discusses particularly the political-religious history of Shi'ism, trends in contemporary Iranian Shi'ism and the role and position of women. One of his main points is that Iranian progressive have not always been concerned enough to argue against the conservatives on their own grounds, using, where necessary argument from the Quran and from the Islamic and Iranian past." The modernization of the status of Westernized women, like other kind of modernization in Iran was buttressed almost entirely by Western-styled arguments and an aping of Western fads likely to shock the traditional." The traditional classes increasingly rejected the whole complex of Westernization (for them tied up with imperialism, the regime of the Shah etc.). The author concludes that whatever the future of the current government, the "ideologies and the feelings they represent must be taken seriously."

161. RAMAZANI (Rouhollah K). Iran: The "Islamic Cultural Revolution". *In* Philip H Stoddard and others, *Ed. Change and the Muslim World.* New York, Syracuse Univ. Pr. 1981, P. 40-48.

What is the meaning of this cultural revolution? Who launched it? Was it a merely a device to use Islam for political ends, or was it a new, genuine attempt to relate the Islamic ideology of the revolutionary regime to the totality of Iranian culture? These are the question which author had tried to explain in the present article.

ISLAMIC FUNDAMENTALISM *in relation to* SOCIETY, MALAYSIA

162. BARRACLOUGH (Siman) Managing the challenges of Islamic revival in Malaysia: A regime perspective. *Asian Survey*; 23, 8; 1983; 958-75.

Author says "One of the most significant social phenomenon in Malaysia in the past decade has been a growing Islamic revival. This study does not seek to explain the origins of these changing attitudes towards the role of Islam in Malaysia or analyse their overall social and cultural significances. Rather this paper has two principle aim. The first is to identify, some of the social, economic and, in particular, political challenges inherant in Malaysia's Islamic revival. The second aim is to analyse the ways in which the national leadership has responded to these challenges and to assess the effectiveness of the strategies adopted to manage the impact of Islamic revival upon the complexities of Malaysian society."

163. NAGATA (Judith). Religious ideology and social change: The Islamic revival in Malaysia. *Pacific Aff*; 53, 3; 1980; 405-39.

Author points out that "The Islamic revival can be seen both as an international movement and one which has specific implications for individual Muslim countries with different meanings in local situation. What follows is a preliminary review of how the Islamic revival has impressed itself on the Muslim population of Malaysia and become an integral part of the political, economic, ethnic, linguistic and cultural scene. Over the past ten years in particular, Islam has been both and agent and symbol of many rapid social changes now occurring on the peninsula. Not only has religion become a source of identity for various elements in Malaysian society, distinguishing Malays and non-Malays, but it also lies the centre of a crisis of legitimacy now emerging among the various elites of Malay society. The author not only described the scope of the Islamic revival, its social and political background, Dakwah movements, political response to Dakwah, its legitimacy of authority but also many other points relevant to the Islamic revival in Malaysia in the present article.

ISLAMIC FUNDAMENTALISM *in relation to* SOCIETY, SOUTH EAST ASIA

164. JOHNS (A H). From coastal settlement to Islamic school and city: Islamization in Sumatra, the Malay peninsula and Jawa. *Hamdard Islamicus*; 4, 4; 1981, Winter; 3-28.

This paper has attempted, with a broad brush, to sketch the outlines of an approach to Islam and Islamic fundamentalism; in particular, in areas which are now part of the parvenu structures called Indonesia and Malaysia. It attempts to look away from Islam with the anthropological, political and economic forces so beloved of the social sciences, and to present it through its religious character establishing its communities through its schools and teachers, its mysticism and its law. The author also discusses the measures for the purification of Islam from corrupting influences and practices; the reformation of Muslim higher education; the reformulation of Islamic doctrine in the light of modern thought; and the defence of Islam.

ISLAMIC FUNDAMENTALISM *in relation to* SOCIETY, SRILANKA

165. SAMARAWEERA (Vivaya). Some sociological aspects of the Muslim revivalism in Sri Lanka. *Soc Compass*; 25, 3-4; 1978; 465-75.

Forming about six percent of the total population, Muslims constitute one of the significant minority social groups in contemporary Sri Lanka. Quite clearly their numbers alone does not enable them to occupy a conspicuous place in society. There are other reasons. Among these perhaps the most important is the dominant position they have traditionally held in trade and commerce; the Muslim trader has always been a ubiquitous figure in the country. Equally important is the role they have played in independent Sri Lanka's political process; their wide dispersion as well as particular concentration in select localities have provided them with an opportunity of acquiring a significant indeed, even a decisive-voice in the electoral contests which have been carried out by national political parties. Despite this visibility, very little ethnographical or sociological information is available about the Muslims - in fact, they have been virtually ignored by scholars working on Sri Lanka. This article attempts to fill the lacunas in our knowledge of the community by focussing on a crucial question which necessarily has to be investigated for a proper understanding of the role the Muslims play in contemporary Sri Lanka's plural society? The thesis which is posited here is that the "revivalist" movement which took place among the Muslims at the turn of the nineteenth century was the major force that moulded their self identity.

ISLAMIC FUNDAMENTALISM, IRAN

166. ARANI (Sharif). Revolution in Iran. *New Quest*; 22; 1980; 201-20.

This article is an analysis, by an Iranian in exile, of the Islamic revolution in Iran. In view of the rising tide of Islamic fundamentalism in various countries this article should prove of special interest for the persons who are specially interest in Iran. The author discussed the circumstances from pre-Islamic revolution of Iran to the establishment of an Islamic theocracy by Ayatollah Khomeini. The author also describes the influence of the Mullah's and their control of the state in Islamic republic of Iran.

167. BRASWELL (George W). Iran and Islam. *Theology Today*; 36, 4; 1980; 523-33.

Author first points out to several movements sweeping across the Middle East, playing havoc with the religious-political dimension (secularisation, geo-politics and resurgence of ethnic-religious minorities such as the Kurds), and sketches briefly Iran's rich religious history, which offers a mosaic of the major religions of the Middle East, "Co-existing in a country which is monolithically Muslim but which has allowed a pluralism of religious communities and freedom of beliefs and practices within ethnic religious categories." The author concludes, however, that the religious revival and political turmoil and revolution in 1978-79 which saw the overthrow of the Shah and the elevation of Ayatollah Khomeini have been formenting for the post decade. Religion and politics in Iran offer an intriguing exploration into the concepts and practices of theocracy, religious pluralism, and freedom of religion." "The treatment of religious pluralism under the present government of Iran remains to be seen. Traditional Islam considers Jews, Zoroastrians and Christians as people of the Book, as compatriots of the revelatory experience from the same God." It is the author's opinion that these ethnic-religious minorities will be controlled in their political influences and participation. "Certainly the Bahai' community will come under closer scrutiny and restriction. And Western mission activities will be approached by the Iranian government in light of its own geo-political concerns. The entire civil religion complex of the Shah will be dismantled."

168. KHUNDMIRI (Syed Alam). Minbar as the symbol of Islamic revival: A critical study of Ali Shariati's radicalism. *In* Anwar Moazzam, *Ed. Islam and contemporary Muslim World*. New Delhi, Light and Life, 1981, P. 61-69.

Author says that the phrase the 'World of Islam' or 'Islamic World' carries an amount of ambiguity. The author poses question whether "we consider changes in Afghanistan too as important or significant in the context of the world of Islam as changes in Iran or Pakistan are considered? "While the modernist does not know what Islam is and defends what he does not know, the revivalist or the conservatives rejects the contemporary consciousness and knowledge of which he has no clear awareness. The author discuss the radicalism of Ali Shariati in this article and also about the importance of the 'Mimbar' as a symbol of Islamic revival.

169. KIELSTRA (Nico). Nature of Islamic revolutions. *Soc Scientists*; 8, 10; 1980, May; 3-21.

Author is of the opinion that while discussing the development in various countries under the name of Islamic revolution, "We have to answer two questions; (1) Why are revolutionary developments taking place at all in these countries? (2) What is the significance of the Islamic aspect of these revolution, and why do we find in Islam a connection between religious fundamentalism and political radicalism that is absent in the other present-day major world religions?" In this paper an attempt is made to analyse the development in Iran because it is the most typical example of an Islamic revolution, together with a brief look at developments in a number of other Muslim countries. The earlier part of this paper based on detailed socio-political analysis and the latter part is frankly speculative and open to discussion. The recent Islamic revival needs not only to be explained in terms of socio-political context in which Islam may be called upon as a political ideology, but we should also try to explain the persistence of the vitality of Islam as a religious conviction.

170. MEHROTRA (O N). Re-Islamization of Iran. *Strategic Analysis*; 3, 5; 1979, Aug: 182-85.

"The recent success of the Islamic revolution in Iran has strengthened forces supporting Islamic fundamentalism and given a set back to modern politico-economic institutions," says author. The various forces joined together under the banner of religion to dislodge a despot in Iran. "Many groups and sections of the people have been insulted from political mainstream of the country. Puritanical Islamic fundamentalist forces might over come the present weak opposition, but this will lead to discontentment and fragmentation. In the long run it might lead to secessionist tendencies in minority ethnic groups", says author.

171. RAY (Baren). Essence of the conflict. *Link*; 23, 48; 1981, July, 12; 16-17.

"The tussle between former President Bani Sadr and the Islamic Republican Party in Iran has often been described by the media as a struggle for power between the secular liberals and the Muslim fundamentalists, but that misses out completely the essence of the conflict." In this article the author has tried to give explanation and the essence of the conflict.

172. RONDOT (Pierre). Islam's protest in Iran: The religious dimension. *Month*; 240, 1338; 1979; 80-85.

The current crisis in Iran has focussed attention on the Shi'ite branch of Islam and its Ayatollahs. Shi'ism has always had a certain relation to political power, and this sets in a religious context the revolt of those who had reached the conviction that the Pahlavi dynasty in Iran threatened the nation with social, moral and political ruin.

The author describes Shi'ism by summerising its history, with particular attention to Husain's martyrdom (*Kerbelah*). "The drama of *Kerbelah* has made a profound impression on Shi'ism whose religious sensibility, exalted for all time by this semi-divine sacrifice, has inspired the loftiest mystical traditions, just as it makes havoc of the emotions of the masses. Sunnism knows nothing comparable." "The essential role of the *moujtahid*, the interpreters who supply for the Imam, is not to be reduced to a political or constitutional initiative: it is a censorship of government activity. The originality of the present movement is that, according to the spirit of the times, this censuring mobilizes the masses of the people and arouses a revolutionary clan which looks for a leader. Thus the Ayotollah Khomeini is incited to assume functions very different from those assigned to him by tradition".

173. SIDDIQI (Kalim). Islamic revolution: Achievements, Obstacles and goals. *In* Kalim Siddiqi and others, *Ed. Islamic Revolution*. London, Open Pr. 1980, P. 9-22.

Author has tried to explain exactly what is involved in an Islamic revolution, what is Islamic revolution. He has also attempted to identify and describe the elements that have gone to make the Islamic revolution. Author is of the view that "the Islamic revolution in Iran is the first defeat of the Western civilization at the hands of Islam." He further says that "this you can see in the reaction of the Western civilization to the triumph of the Islamic revolution in Iran." "This revolution has given us hope, says author, where there was no hope. This Revolution has restored confidence in us."

ISLAMIC FUNDAMENTALISM, IRAN, HISTORY

174. ABIDI (A H H). Islamic revival in Iran. *Seminar*; 290; 1983, Oct; 25-29.

In this article the author says that 'Islamic revivalism is not new to Iran and traced the developments from Safavid rulers to present century. A radical shift occurred with the advent of Ayotollah Khomeini in 1963. He openly opposed monarchy and advocated that those trained in religious law should rule the country. The author says that "Islamic resurgence has to be understood in the context of the distinct religious doctrines of Shi'i Islam and described them in length. " This Islamic resurgence in Iran is not a freak phenomenon. It is the product primarily of long continuing religiousness of the people gaining a political expression in the face of an alienated, unresponsive, and current political and administrative system." But "Islamic revivalism as propounded by Khomeini and his kind of divines has created a host of problems for the people and, in turn, revivalism confronts numerous challenges that come from many directions." "The Iranian regime has launched an all out-cultural revolution in order to reach its cherished goal of Islamization. The political dimensions of Islamic resurgence in Iran have had an impact on the region and also to neighbourhood where the socio-political and military strategies are being revamped."

175. ASARIA (Iqbal). Iran-a case study in Muslim political awakening. *In* Kalim Siddiqui and others, *Ed. Islamic Revolution* London, Open Pr. 1980, p. 23-36.

According to author "the roots of the Islamic movement in Iran lie firmly in the Ulama's uncompromising stance against tyranny and foreign domination and their exercise of their role as guardians of religion during the *ghayba* of the Imam. The movement has gained experience as it has unfolded, in particular it has now gained sufficient strength to dispense with the alliance of 'liberals', secularist and leftists of various colourations. The opportunistic behaviour of all these groupings and their inherent unreliability has led to a formulation for a demand for the establishment of an Islamic Republic led and guided by the Ulema; all other organisations having to subsume their aims to the Islamic movement." "The Islamic movement in Iran is coloured by the history and background of the Iranian Muslims, but in its broad sweep it is part of the world Islamic movement and derives its strength from the universal desire of the *Ummah* to implement unadultrated Islam," says author.

176. GRIFFITH (W E). Revival of Islamic fundamentalism: The case of Iran. *Hamdard Islamicus*: 3, 1; 1980; 47-59.

The author says that "the Iranian revolution has high lighted one of the principle religious and political development of our time, the revival of Islamic fundamentalism, from Indonesia to Morocco and from Turkey to Central Africa. In the short run it will cause more problems to the West. In the short run, however, it may be more dangerous to the Soviet Union in Muslim Soviet Central Asia." "The revolt against western style modernization is today especially strong in the Islamic world." There have been recurrent waves of Islamic fundamentalism in Muslim countries. Among Middle Eastern countries, Iran has two major political traditions, antique Persian Kinship and Shi'i Islam. At the beginning of the sixteenth century the Safavid dynasty made Shi'i Islam the state religion of Iran. The Shi'i doctrine produced a series of Iranian millenarian movements. In 1963, Khomeini was exiled from Iran for his protest against the Shah's autocratic rule, and his close ties with the United States and Israel — the same protest he made in 1978-79 which brought Islamic revival in Iran. There have been many other causes of the Iranian revolution besides Islam, which are described in this article in short.

ISLAMIC FUNDAMENTALISM. IRAN *in relation to* POLITICS, INTERNATIONAL

177. RIZVI (S A A). Iranian revolution and the Islamic revivalism. *Hamdard Islamicus*; 5, 4; 1982; 89-108.

Author disagreed with the picture of the Iranian and Islamic revivalism as a whole such as is found for instance in G H Jensen's 'Militant Islam' and similar publications, which seem attractive to the Western mind but they provide little perception of recent developments in the Islamic world and contemporary society." The author attempts to provide a deeper perception of the process "involved in its success and in its impact on both the world of Islam and the West respectively," by an extensive historical analysis and an assessment of the recent political development. He concludes, for instance, that to Ayatullah Khomeini the Islamic republic meant the translation of the revolutionary ideals of the Quran and Muhammad, fighting the imperialistic designs of super powers," "the Iranian revolution was brought about by the *musta'ifin* (deprived, oppressed, exploited or poor) and would be defended by them." The oil-producing countries "adhere to the elicited loyalty from their subjects only by brute military force." The Iranian revolution and Islamic revivalism are pitted against imperialism and colonialism without prejudice to scientific and industrial developments.

ISLAMIC FUNDAMENTALISM, ISLAMIC ORDER

178. GEYBELS (M). Salient features of an Islamic order. *al-Mushir*; 20, 3; 1978; 90-98.

Starting point is Pakistan's present design of an Islamic order (Nizam-i-Islam). The Prophet Muhammad "had a great deal to do with the establishment and organization of the Islamic community (Umma) and its order (Nizam)." "If we wish to understand the nature and features of Islamic order, it is necessary to study the condition of Arabian society as they were at the time of Muhammad." The origin of the new community and its order were primarily based on religious loyalties and considerations which surpassed all other. The author further deals with "special characteristics of an Islamic order," the definition and consequences of an Islamic order, and its implementation. By way of conclusion, the author refers to Mohd. Asad (The principles of state and government in Islam, Los Angeles, 1961) who wrote in 1961 that "none of the existing Muslim countries has so far achieved a form of government that could be termed Islamic." "If this is true, Pakistan will be the first country to achieve this goal, when the Nizam-i-Islam has been introduced fully.

179. KRAMER (Martin). Ideals of an Islamic order. *Washington Quart*; 3, 1; 1980, Winter; 3-13.

A composite Islamic order provides for a state integrated with religion, a state selfavowedly Muslim, led by Muslims, in which the law of Islam enjoys a privileged position. The political system is said to follow Islamic precedent, the economic system is held to exemplify Islamic values of social justice. Society is aligned with the state's interpretation of Islamic principles, and the state enforces that interpretation through state institutions. Its foreign policy will reflect a special obligation towards other Muslim state and people.

180. YUSUF JALIL. Political aspects of an Islamic order. *al-Mushir*; 20, 3; 1978; 99-107.

Describes the salient feature of the first Islamic state as it was established by Muhammad. The subjects dealt with are; theocracy, Islamic nationalism, the idea of 'one nation', God given authority for the head of the state, justice, the holy war, consultation, different departments in the state (instituted by Muhammad), state religion, state policy, administration. The author further mentions a number of characteristics of the state ruled by the rightly-guided caliphs (632-661). It is concluded that the Quran and the Sunna give only very general principles of government and indicate more the spirit with which a government should be exercised than its political theory. There are, however, "at least some characteristics and qualities an Islamic rule ought to manifest. Such are: an Islamic state is a theocracy; as based on God's law (Sharia), the Islamic state ought to separate strictly the legislative and the executive powers; as the Islamic state is "egalitarian". providing complete equality to all citizens, no citizen has any spiritual or legal ascendency over another one; an essential condition for any Islamic government is *Shura* or consultation between the ruler(s) and the ruled. This is the only clear principle in the Quran, but it is very emphatically proclaimed. In this sense Islam can be called "democratic."

ISLAMIC FUNDAMENTALISM, ISLAMIC ORDER, LAWS

181. DASKAWIE (M A Qayyum). Legal aspects of an Islamic order. *al-Mushir*; 20, 3; 1978; 108-114.

"Since Islam is not only a religion but also a state, its fundamental character is not really a theocracy but a democracy." "The Islamic code covers all aspects of the believer's life; as believer; as a man in relationships; and as a citizen of an Islamic state." "Islam (We constantly hear) is a complete code of life (Mukammal zabita-i-hayat)." The author deals with several aspects of this code, such as; *aqaid* or rights of creatures (one's responsibilities towards fellow human beings). Apart from these aspects there exist the legal and juridical aspects of the Islamic state, which are briefly characterised. Finally it is emphasized that the demand for the introduction of the *Sharia* in Pakistan has brought into sharp focus some very important but controversial problems, e.g. the phrase "in the name of Islam" has meant different things to different sections of the people, should all the old existing laws be scrapped? The conservative elements have been clamouring for the introduction of the total *Sharia* immediately as a panacca for the many ills of the society. There are others, however, who while agreeing with this demand would like to see first that the other conditions (social, economic and cultural) as are provided for in the *Sharia* are fulfilled as a pre-condition to the imposition of the *Sharia*. These conditions, they claims, existed in the ideal Islamic state of the first four "rightly guided Caliphs."

ISLAMIC FUNDAMENTALISM, ISLAMIC ORDER, STATE

182. ANWAR MOAZZAM. Resurgence of Islam: Role of the Peoples and the state. *In* Anwar Moazzam, *Ed. Islam and contemporary Muslim world*. New Delhi, Light and life, 1981, P. 1-10.

The expression "Islamic resurgence" has been given currency chiefly by the Western media, apparently prompted by the Iranian Revolution, Pakistan's adoption of Nizam-e-Mistafa and the confident and blunt statement of Qaddafi of Libya on Islamic reconstruction of Muslim Societies. Author says that "if by this term is meant the strengthening of Islamic fundamentalism, then it is not at all a new trend. What is new is the state sponsored Islamic revival as a political and social system through partial implementation of *Sharia*. There have been consistent organised efforts at the theological and social levels to purify Muslim minds and society. In India certain thinkers like Shaikh Ahmad Sirhindi, Shah Waliullah, Tahrik-e-Mujahidin of Syed Ahmad Barreli, Abul Ala Maududi and present day organization like Tablighi Jama'at and Jama'at-e-Islami, in Egypt and Arab world, the Salfia movement, Ikhwanul Muslimoon, and the revivalist movements in Malaysia and Indonesia, representing long tradition of Islamic fundamentalism. The medieval image of Islamic social system has to tackle with the world of 20th century creating the problem of establishing a co-relation between the eternal values of Islam and the changing societies. The author says that "the present Muslim intellectual leadership is still divided into two camps-one traditional-bound and the other refusing to be tied down to tradition. It is to be noted, however, that fundamentalism in the sense of recognition of the Quran and Hadith as the basic sources of guidance, is not a feature of Muslim orthodoxy exclusively; the modernist like Afghani are also fundamentalist in this sense. ''According to author the "state-sponsored phenomenon of Islamic resurgence" is the only method through which Islamic system can be given a concrete shape. The author poses a question that "there are at present about 15 Muslim countries in the world proclaiming Islam as the religion of state. How many of them have been able to implement *Sharia* as the law of state and that too to what extent"? But "certain Muslim states are trying to bring the Islamic legal code in tune with the demands of modern society. In the circumstances, it is premature to identify the direction the Muslim culture would take.

183. ENGINEER (Asghar Ali). Religion and the nation state. *Seminar*; 290; 1983, Oct; 30-32.

Author opines that "Islam and nationalism are often considered contradictory terms. Apart from theologians, many other Muslims trained in modern disciplines also maintains that Islamic internationalism is both political as well as religious. "Having recognized all the important determinants of *qaumiyyet* (Nationhood) i.e. language, culture and territory it would not be in the right spirit of Koran to maintain that there is not place for nationalism in Islam. It is very difficult to reject in practice the reality of the modern concept of nation state. The Jamaat-i-Islami believes in religious nationalism if it can be so described. It believes that Islam is a sufficiently strong religious bond to make for a viable unity and to constitute a state. The recent Iranian Revolution has once again given spurt to the idea of Islamic internationalism. But certainly "no Arab country, however, rich in its oil resources, view with equanimity the prospects of Muslims from other countries settling down and sharing their riches. The author also says that "the confrontation between Arab nationalism and Islamic internationalism has been a predominant feature throughout the Arab world, including Saudi Arabia which provides leadership to the conservative Arab regimes."

184. FAKSH (Mahmud A). Basic characteristics of an Islamic state. *J South Asian Mid East Stud*; 5,2; 1981, Winter; 3-16.

The world is witnessing a new effervescence of Islam that is manifest in the current global "recovery of Islamic identity." This Islamic recovery and the events in Iran that culminated in the success of the Islamic revolution under Ayotollah Khomeini in 1979 have once again brought to focus the concept of an Islamic state. The question of an Islamic state based on Islamic constitutional principles has become a major concern to most political and religious thinkers in the Muslim world. To reach a better understanding of this critical issue, this study attempts to examine and explore in depth the question of the basic characteristics of the Islamic state.

185. GELLNER (Ernest). State and revolution in Islam. *Millennium*; 8, 3; 1979-80, Winter; 185-99.

The author examines the present political validity of Islamic societies by a sociological analysis of the historical development of Islam and its link with the rise of the modern state. He argues that religion plays a triple role in modern Islamic states in confirming a continuing historical tradition, by rejecting the foreigner, but by also providing a norm of self-discipline and social purification.

186. al-GHAZALI (Muhammad). Introduction to a draft Islamic Constitution. *Islamic Stud*; 20, 2; 1981; 153-68.

One of the most challenging questions in the contemporary Muslim world is that pertaining to the form of government, the viability and practicability of Islam as a foundation on which the structure of a modern, democratic progressive, just, egalitarian and welfare state can rest. Many studies have been and are undertaken on the political theory of Islam and, the latter being compared with other contemporary ideas. The Islamic Research Academy of Al-Azhar appointed a committee of its scholars to work on the Islamic theory of state and its application in the modern times. This committee has recently brought out a draft Islamic constitution, Imam, his election, fundamental rights and civil liberties, broad lines of state's economic policy, the concept of Umma and its reconciliation with the plurality of state etc., but carefully avoiding going into any controversial detail leaving its settlement to the *Shurah* stipulated in the constitution. The author finds its values and worth in the fact that it provides a framework which gives a scope for further discussion among Muslim scholars and exponents of political thought of Islam. The draft of this constitution is published here, in a translation from Arabic by the present author.

187. HAMOODUR RAHMAN. Islamic concept of state. *Hamdard Islamicus*; 2, 1; 1979; 51-64.

Author says "that we have advanced from the stage of merely talking about the establishment of the Nizam-i-Islam in Pakistan". Now "it has become necessary for us to ascertain as to what is the true form, character and nature of an Islamic state and how far it can be realized in the structural pattern of a modern society." Now the author says that " the first Islamic state was set up by the Prophet himself." The main characteristics of this state are that this institution should be establish through a democratic process. That it should have a constitutional organization functioning under a written charter. That it should have a federal structure. That it should be based on the concept and fundamental principles embodies in the Quran. The following are the fundamental principles - The concept of human sovereignty is completely absent in the political philosophy of Islam. The principle of Tawhid is the fundamental principle of Islam and the *Kalima Tawhid* is the greatest charter of individual liberty. Islam seeks to set up a just society and therefore, attaches the greatest importance to justice, equity and fair dealing - the affairs of the state were not only required to be regulated in accordance with the laws of the Quran but also to be administered through mutual Consultation - Authority of power to rule according to Islam is a trust - *Amanat* - of the people and not the birth right of any one. This was indeed a unique welfare state for which no parallel can be found in the history of the world. The author has given only a bare outline of the pattern that it may be possible to adopt in the light of modern condition in order to bring our system of government into line, as far as possible, with the Islamic pattern keeping in mind its essential characteristics.

ISLAMIC FUNDAMENTALISM, ISLAMIC ORDER, STATE, CRITIQUE

188. TASNIM AHMED (S). Myth of Islamic State. *Mainstream*; 41, 12; 1982, Nov. 20; 17-20.

The concept of Islamic state is a myth—a figment of un-tamed imaginations - which.results from a misunderstand-ing and mixing up of the Prophet's two fold role in the later years of his life. Apart from being politically mischievous and misguided, the concept only belittles the grand mission the Prophet had set out to perform - true to his stature and station, he worked all his life for the creation of a noble world, and not to establish a Muslim emirate, what to speak of a 'mythical Islamic state'.

ISLAMIC FUNDAMENTALISM, ISLAMIC ORDERS, STATE *in relation to* DEMOCRACY

189   JAVID IQBAL. Democracy and the Modern Islamic State. *In* John L Esposito, *Ed. Voices of resurgent Islam*, New York Oxford Univ. Pr. 1983, P 252-60.

A fundamental issue underlying attempts to establish modern Islamic states is the nature of authority and the role of democracy in Islam. The author says while Islam is open to the development of new political and legal systems, the ideological commitment of the state does set certain limits or restrictions affecting such areas as candidates qualifica-tion for elective office, the conduct of elections, the legis-lative scopes of a national assembly or parliament, and the political party system "only such political party which ad-here to Islamic ideology can be permitted to function in a Muslim national state."

ISLAMIC FUNDAMENTALISM, ISLAMIC ORDERS, STATE, IRAN

190. ARJOMAND (s). State and Khomeini's Islamic order. *Iranian Stud*; 13, 1-4; 1980; 147-64.

"In Khomeini's Islamic order, the state has not only been constitutionally weakened but also made 'Islamic' i.e. manned, in so far as possible, by reliable Islamic personnel and brought under direct or indirect clerical control." This article is the expression of the above feelings of the author where he discusses state and Khomeini's Islamic order.

191. FERDOWS (Amir H). Khomeini and Fadayan's society and Politics. *Int J Mid East Stud*; 15, 2; 1983; 241-257.

Under the titular leadership of Ayatollah Khomeini, Iran is struggling to establish an Islamic republic. Such a system of government based on an Islamic constitution and under pinned by Islamic laws which are 'all encompassing and complete, is strongly believed by him and his followers to be the Panacca for all the socio-economic and political ills of Iran and, for that matter the whole Islamic world. A few decades ago, the movement of Fadayan-i-Islam advocated a strikingly similar system of government, with only a few shifts in emphasis. This religio-political movement tried to undo what had taken place in Iran in the way of modernization and westernization since the 1920's. There goal was to replace the Shah's government with an Islamic regime under an Imam." ....."The Fadayan have a headquarter now with Ayatollah Khalkhali as the self proclaimed chief."

This article presents a comparative analysis of Khomeini's ideology with that of the Fadayan.

192. LAKE (C M). Problems encountered in establishing an Islamic Republic in Iran, 1979-1981. *Brit Society Mid East Stud Bull*; 9, 2; 1982; 141-70.

This article tries to ascertain "why Iran's post-revolutionary leaders have had to resort to oppression in order to establish their government of righteousness." A study is made of the attempts to apply the 1979 constitution to Iran from 1979 to 1981 as well as a study of the constitution itself. The examination is divided into two areas, theory and practice. The former consists of an examination of the constitution and evaluates whether it can be applied theoretically in Iran. The empirical side deals with the efforts by the *Ulama* to apply the document to Iran from 1979 to 1981 and analyses the method of government in Iran during this period. The article further analyses the role of *Mullas* in the revolution of 1978-79 and evaluates why this was to be of significance in the post-revolutionary phase. The study concludes with attempts at an 'explanation of the problem in Iran since 1979 by sources in, or close to, the Islamic government" (examined here are three views of the causes to the failure of the Islamic Republic).

193. MALIK (Hafeez). Emerges of an Islamic State in Iran. *J South Asian Mid East Stud*; 2, 3; 1979; 3-5.

Author says that "while we bid farewell to the Shah, we welcome the leaders of the new Islamic Republic of Iran. What is an Islamic state? In the west, most people do not have any understanding of this polity. In popular imagination, some of the harsh penalties provided in the Islamic penal code are equated with the Islamic order. Not many people realize that an Islamic state paradigm establishes executive, legislative and judicial branches of the government separately. Rights of citizens and non-Muslims minorities are carefully enumerated and protected." "Muslim scholars also have an Islamic view of international relations." "The world would understand the Islamic state in its living reality." "Ayatollah Ruhollah Khomeini's leadership in Iran now bears a special responsibility in projecting a new Islamic order which will be just an egalitarian, and ensure basic fundamental rights to all citizens."

ISLAMIC FUNDAMENTALISM, ISLAMIC ORDER, STATE, LAWS, LIBYA

194. TAHIR MAHMOOD. Reflorencence of Islamic law in Libya. *In* Anwar Moazzam, *Ed. Islam and contemporary Muslim world*. New Delhi, Light and life, 1981, P. 143-58.

"The re-orientation of laws in Libya in pursuance of its cultural revolution is indicative of an attempt to enforce traditional jurisprudence in a modern environment. While introducing the Quranic penalties for the offences of theft, decoity etc." The re-Islamization of the Libyan civil and commercial laws has not been as striking as that of the penal code. From the very beginning, Islam had a well developed system of civil and commercial legal principles. More particularly, the labour laws of Islam had a well developed system of civil and commercial legal principles. More particularly, the labour laws of Islam always had an amazing touch of modernity. Libya has now enforced both the civil and the penal laws of Islam. By the "Islamic laws" which is being re-introducing in Libya is, it is evident, meant the broad, flexible frame work of the *Sharia* and *fiqh*, which admit a wide latitude for selection, deduction and also formulation of new legal rules. Proud of its ultra-modern jurisprudence and legislation, the West might wink at the development in Libya. The gran re-enforcement of the Islamic legal system may embarrass those orientalist who always take pride in locating the trends of westernization in the East.

ISLAMIC FUNDAMENTALISM, ISLAMIC ORDER, STATE, LAWS, PAKISTAN

195. MOHAMMED SAID. Enforcement of Islamic laws in Pakistan. *Hamdard Islamicus*; 2,2; 1979; 61-90.

Author in this paper quotes a tradition of the Prophet of Islam, where in the Prophet gives the Islamic state a status of a trustee to protect and preserve faith. Muslim league promise to make Pakistan a state where Islamic law would be supreme. "However we observe that what we expected about the enforcement of Islamic Sharia in Pakistan and what we anticipated about this independent Islamic state to be a model of Islamic revival in the 20th century was all probably mere imagination or a mirage." "During this period the position of Islam in this very Islamic state had become more like an orphan who was always exploited to the advantage of the rulers while itself being deprived of every privilege. Now the announcement of the President of Pakistan and his determination to Islamise the laws proves that Nizam-i-Islam is not only practical but also an imperative requirement for the man of this age. The author says "It makes me still more happy when I see that Pakistan is not alone in its determination to enforce Nizam-i-Islam."

ISLAMIC FUNDAMENTALISM, ISLAMIC ORDER, STATE, LAWS, SHARIAT, PAKISTAN

196. TANZIL UR REHMAN. Pakistan on road to Islamisation. *Radiance*; 17, 23; 1981, Oct., 18; 3.

Describes how the government of Pakistan on the recommendation of the Council of Islamic ideology, took steps in the implementation of *Sharia* which will have far-reaching effects on the reformation of the society. The council recommended a three phased programme for the elimination of interest from country's economy. If efforts in Pakistan succeed, it will pave the way for Islamization of banking and economy in other Muslim countries. In the field of education the Sharia Faculty was established in universities. Islamic tradition were introduced in judiciary and in other fields. The author is of the opinion that "with all this it can be rightly ascertained that the step taken and to be taken in Pakistan, in implementation of Sharia will go a long way, and the author believes, that within the next two or three years, every one of us will feel a visible change in the socio-economic, educational and political spheres of life of the Muslim society in Pakistan."

FUNDAMENTALISM, ISLAMIC ORDER, STATE, PAKISTAN

197. DAND (Satya Pal). State and religion: A lesson from Pakistan. *Mainstream*; 19, 42; 1981, Jan, 20; 29-33.

Pakistan was established on 14th August 1947, as an Islamic Federal Democratic state. In thirty years of the existence of Pakistan it broke up into two countries - Pakistan and Bangladesh. Present military dictator of Pakistan Ziaul Haq has categorically declared that there will be no election in the near future, that no election but the establishment of Nizam-e-Islam in the country is the primary task. He now promises transfer of power to the people only when "the objective of Islamization and revival of Islamic values" has been achieved. This Islamization is being achieved through Islamic financial and tax system such as introduction of *Zakat*. Another important measure to Islamize the economy was the introduction of interest free banking. The result of these measure was that Nizam-e-Mustafa of Zia has been characterised as un-Islamic by many important political and religious leaders in Pakistan. They said that "Islam does not permit one man rule or dictatorship and so the General's government cannot be considered as Islamic. Such have been the results of Islamization of Pakistan carried out on the basis of the theory that Muslim are a nation and that religion and politics are inseparable. The author quoted a paragraph from Patriot (June 4, 1981) in which Chief Minister of Baluchistan has drawn from experience "Look, I want to reiterate that there is no commodity like an Islamic nation. It has never existed and it cannot be artificially created in the name of our holy scripture - The Quran. Those who take of an Islamic nation or a Muslim nation are committing a heinous crime against the various nations and nationalities which subscribe to the Muslim religion."

198. DE VEGA (G C).Pakistan as an Islamic State. *Asian Stud*; 6, 3; 1968, Dec., 263-70.

Analyses and interprets the meaning of Islamic State in Pakistan. Concludes that Pakistan, for all practical intents and purposes, still continues to be an Islamic state not withstanding the controversial honing of its previous appellation Reliance on Islam in the preambles of both the constitutions a preponderance of belief and faith in the force and validity of Islam manifested in various ways by the general body of the Pakistani Muslim population, description of Islam as not only a religion but also 'a way of life' etc. are additional aspect dealt with.

199. ESPOSITO (JOHN L). Pakistan: Quest for Islamic identity. *In* John L Esposito. *Ed. Islam and development.* New York, Syracuse Univ. Pr. 1980, P. 139-62.

Examine Pakistan's attempt to translate her Islamic aspiration into political and social reality. In the 1970s Islam re-emerged as a major component in Pakistan's political development. A commitment to the introduction and enforcement of an Islamic system (Nizam-e-Islam) has become the chief means of the present ruler. General Ziaul Haq to legitimate his coup and confined rule. Islamic political legal, and social reforms have been introduced. Further more, Pakistan's example may have implications for other Muslim countries who seek greater Islamization for their political and social systems.

200. JAURA (Ramesh). Islamization of an Islamic republic. *Econ and Pol Wkly*; 14, 24; 1979, Jun, 9; 966-67.

In this article the author says that "on February 10, 1979 (12th Rabi ul Awwal, the birthday of Prophet Mohammad) Pakistan's Chief Martial Law Administrator and President Ziaul Haq announced the much-heralded Islamization measures. The measures deal broadly with taxation and deterrent punishments for various crimes. Such measures to 'Islamicise' the Islamic Republic of Pakistan where necessary to justify the regimes existence. Perhaps the intention of these Islamic measures is not to reform the present system but only give the appearance of doing so.

201. MEHROTRA (O N). New measures for Islamization of Pakistan. *Strategic Analysis*; 4, 4; 1980, July; 141-46.

The measures which claimed to be a step towards implementation of Islamic precepts, enabling Muslims to order their lives, in the individual and collective spheres, in accordance with the tenets of Islam were imposed by an ordinance of General Zia-ul-Haq on 20th June, 1980. The ordinance has been extended to the whole of Pakistan and applies only to Muslims, says author. The new economic measures for Islamization for Pakistan were introduced. Islamic panel code were also enforced. The purpose of implementation of Zakat and *Ushr* ordinance is believed to be to reduce social disparities and imbalances and is said to constitute an important first step towards social justice as conceived in the Quran. But after the implementation of these measures various serious controversies have been raised. In this present article the author has clearly proved that Islam in Pakistan is not a monolithic force. Muslims in Pakistan will taxed differently. It is not clear whether this will increase the popularity and following of the *Shias* who will now be exempt from compulsory *Zakat* and all the bureaucratic intrusion it involves. The rulers in Iran who are Shia fundamentalist and who have been competing with Sunni fundamentalists else where in propagating their brand of Islamic faith are bound to extend their support and sympathy to their brethren in Pakistan. Author concludes "as efforts are undertaken to resort to religions sectarian schools of beliefs and emphasise the differences among them. This has been the lesson of history and the recent events in Pakistan once more validate it."

202. MEHROTRA (O N). Revival of Islamic fundamentalism in Pakistan. *Strategic Analysis*; 3, 6; 1979, Sept.; 214-20.

General Zia-ul-Haq introduced measures for the Islamization in Pakistan. In the process he was easily won over by Islamic fundamentalists and promised to introduce Islamic laws and through them cleans the administration. In this endeavour, he was supported by the conservative, orthodox, rightist parties. However, the introduction of Islamic laws has not strengthened the position of President Zia. More over introduction of Islamic laws have created a host of problems. The author is of the opinion that "under present circumstances, it does not seem possible that religion can solve the problems of Pakistan. The author further stressed that "the whole system has become confusing and beset with contradictions. Present efforts to establish a new politico-economic-military-religious set-up, in all likelihood, would face insurmountable difficulties."

203. MUMTAZ AHMAD. Pakistan: Islam, the military and the

politics of trivia. *Arabia*: 26; 1983, Oct; 12-14.

As far as Islam is concerned it has played a very important role in the country's constitutional disputed, political debates and the socio-economic controversies. Before Zia's rules paid only lip service to Islam. Zia proclaimed for Nizam-e-Mustafa in Pakistan and got support from different corners. Among the Islamic measures introduced so far by Zia the Islamic penal code, *Zakat*, *Ushr*, *Shariah* Courts, prayer breaks during working hours etc. There is hardly anything that threatens the status quo. *Shura* has no legislative power it is only a debating forum. Meanwhile the regime has developed its own religio-political ideology emphasising the need for a continued and effective political role of the military, a limited franchise, controlled press, hand-picked *Shura*, and a lot of Islamic rhetoric.

204 MUNEER AHMAD KHAN. Islamization in a Muslim country - The Pakistan experiment. *In* Anwar Moazzam, *Ed. Islam and Contemporary Muslim World*. New Delhi, Light and life, 1981, P. 71-83.

The move to Islamise Pakistan is not new. Islamisation has however gained momentum and frequency since General Zia-ul-Haq took over from Bhutto on July 1977. The term 'Nizam-e-Mustafa' was popularised by the Pakistan National Alliance which among nine other parties who included the Jamat-e-Islami and the Jamiat-e-Ulama-e-Pakistan. In 1974 Bhutto was compelled to take certain other drastic reformatory steps as a conciliatory move on his part for both domestic peace and appeasement of the fraternal and fundamentalist Muslim countries such as Saudi Arabia and Libya. Gambling and racing were banned and prohibition enforced in Pakistan. The reconstitution of the Advisory Council of Islamic ideology for suggesting the Islamization of the existing laws of the country and the establishment of the *Shariat* benches.

Zakat and *Ushr* in both their voluntary and mandatory forms are being revived to remove economic disparities. *Huddods* are to be imposed to reduce the rate of crime. In addition, changes in the educational system and broadcasting have been introduced. Text books are being revised to bring them in confirmity with the ideology of Pakistan. Unlike Iran, the Zia regime has taken to gradualism. The pace has been slow and selective with no rigidity in matters of the promulgation of Islamic theology. Unlike Iran again, even the demand for the uprooting of the earlier set up has not been total in Pakistan. In fact even today, for a variety of reasons, it is more likely that the parties believing in the enforcement of Islamic law would not muster a sizeable number of seats in any fair election and this despite their fundamentalist approach.

205. RAZVI (Akhtar Adil). Islamic transformation in Pakistan. *Radiance*; 17, 15; 1981, Aug, 23; 3.

The Islamization programme in Pakistan raises several questions: What does it involves? What has been achieved so far, and what are the concrete plans towards this end? Will Pakistan be able to maintain its modern character alongwith the implementation of the *Shariah* in all aspects of its socio-economic and political life? These are the questions which have been raised by the author in this article. The author is of the opinion that the Islamic transformation that Pakistan (and other Muslim countries have taken in hand) is not going back, but moving forward. Because "history bears testimony to the fact that Islam was the cause of healthy revolution in whatever country, region or society it reached. A reassertion of the centrality of Islam is not going back to a lesser state of social or political evolution. It represents, in essence, a forward looking attitude."

ISLAMIC FUNDAMENTALISM, ISLAMIC ORDER, STATE, SOUDI ARABIA

206. MAC INTYRE (Ronald R). Saudi Arabia. *In* Mohammad Ayoob, *Ed. Politics of Islamic reassertion.* New Delhi, Vikas, 1982, P. 9-29.

Rapid technological development and their use had a sharp reaction among *Ulmas* of Saudi Arabia, and we have to see that to what extent does Islam confirm to the dynamics of new religo-modernising synthesis. In the 18th century Hanbalite religious reformers traced to eradicate *Shrink* from Islam. Wahabism was the driving force in the expensive messianic movement. In the beginning of the 19th century Saudi expansionism was curbed by Ottoman Sultans in inter tribal division weakened them from inside. Since the king has all powers in the state, but he is bound by Sharia and all his orders are determined in consultation with Ulama. The growth of oil industry and increase in literacy are the two factors which may oppose the system in the comming period. The opposition fall into two categories: Ideological and revolutionary, fundamentalist and revolutionary. Islam is an instrument of Saudi Arabia foreign policy and was adopted as an ideological defence of internal and external security interest. Two related factors enabled Saudi Arabia to emerge as a dominant middle east state from 1967 onwards, (a) changes in the balance of power in the region, (b) increasing oil wealth during the 1970s. To conclude author says that the stability of Saudi Arabia in the 1980s depends on its continuing ability to adopt its traditional institutions and values to the process of modernising change. To accommodate change and internal stability the Saudis must perfect consensual political institutions. And for an oil hungry world instability in Saudi Arabia seems as awesome prospect.

ISLAMIC FUNDAMENTALISM, ISLAMIC ORDER, STATE, SUDAN

207. MINAI (Ismail Ahmed). Sudan opts for the Islamic way. *Islamic Order*; 5, 4; 1983; 61-72.

The explosion of an Islamic era in Sudan is an event of extra-ordinary significance. A conspiracy of silence by Western Media has sought to under play the tremendous and momentous change. What little has appeared in Press was mostly either a disguised or distorted version of the reality. Impact International has done a great service to project the revolutionary shift in proper perspective. Nimieri and Turabi provide a rare insight about the transition and their thinking also indicates the shape of things to come. For Pakistan, which has also embarked on a similar course and is passing through testing times, the Sudanees experiment and experience of Islamic resurgence has an obvious importance.

208. SALIHEEN (Mohammed Khojali). Islamization in Sudan. *Radiance*; 19, 36; 1984, Jan., 15-21; 4.

Author discussed the recent Islamic measures in Sudan which is the beginning of the application of Islam in various aspect. The author also discussed the reaction of Sudanese mass-media towards recent Islamic legislation and also out come of the Islamic experiment in Sudan.

209. TURABI (Hassan). Islamic State. *In* John L Esposito, *Ed. Voices of resurgent Islam.* New York, Oxford Univ. Pr. 1983, P. 241-51.

Author is founder of the Sudanese Muslim Brotherhood. As he notes, he is "directly involved in the political process that seeks to establish an Islamic State." The author presents some of the essential or universal characteristics of such a state, focussing on its nature and functions There are many issues surrounding the creation of modern Islamic states. How open is Islam to modern nations of state and society: nationalism, popular sovereignty, democracy, freedom and dissent, popular political participation and representation, modern law and economics? The remainder of this section presents selections which delineate the nature and purpose of an Islamic state and society. Given the ideology of an Islamic state, what place is there for nationalist or pan-Islamic identity and allegiance? What are the implications of an Islamic state regarding questions of sovereignty, election, political parties, and the rights of non-Muslims?

ISLAMIC FUNDAMENTALISM, ISLAMIC SOCIALISM

210. ENGINEER (Asghar Ali). Islam and socialism: An interpretative approach. *Islam Mod Age*; 10, 1; 1979; 37-57.

Examines the question of relationship between Islam and modern socialism in the light of socio-historical factors. This examination leads to the following: Today's monopoly capitalism is totally opposed to the egalitarian spirit of Islam. The concentration of wealth and of the means of production negates the rationale of the Islamic value system. In almost all the Muslim countries, however, "the ruling classes are reactionary and thus they collaborate with western imperialism to perpetuate their role. And as a dope to the masses, they talk of establishing Islamic state by enforcing the Islamic penal code (cutting of hands for theft, stoning one to death for adultery etc.) as if it is the most important ingredient of the Islamic State. The concept of equality goes against their class interest and, therefore, this core value which was emphasised most by the Prophet in the earlier period is quitely ignored. It is, therefore, necessary to re-emphasize this core-value of Islam if the Muslim masses and not the Muslim rulers are to benefit from it. One can thus say in conclusion that the value-system of Islam has much in common with socialism although its thought and institutional system do not. And, in an industrial era, in order to realise the fundamental values of Islam, new instrumental values like planning, socialisation of the means of production etc. will have to be adopted."

211. RIZVI (S Ameenul Hasan). Use of Islam for socialist ends. *Radiance*; 19, 26; 1983, Nov. 6-12; 2.

Author says that "with variations in degree and speed, the process of Islamization of the Arab-Muslim world continues. Some countries have openly declared themselves Islamic while in other public pressure has yet to reach the stage where the rulers may not be left with any other alternative to Islam. Egypt, Iraq, Syria, South Yemen, Tunisia and Libya etc. have opted for socialism. There is yet another kind of country like Sudan where the ruling socialist party has used Islam to serve its own socialist ends." "To the Sudanese President Mr. Gaffar Nimeiri, some 'Islamic pretensions; were necessary. Therefore out of socialist necessity Islam has been enforced in the country. But he (Mr. Nimeiri) should have learnt a lesson from Mr. Bhutto's experiment in using Islam for socialist purposes," says author.

ISLAMIC FUNDAMENTALISM, ISLAMIC SOCIALISM, PAKISTAN

212. BAIG (B G). Islamic fundamentalism. *Soc Scientist*; 9, 1; 1980, aug. 58-65.

Islamic fundamentalism can best be understood by analyzing Pakistan's path of development that country has always been found it to be in search for a theoretical foundation. They look for a way in compromise in the reconciliation of capitalism and socialism. The Pakistan version of the theory of middle road is often designed as Islamic socialism. The Islamic socialists often maintain that the purpose of their programme is to cure capitalism of its ills and the latter can be eliminated by following the three Islamic precepts, i.e. *Zakat*, observance of Islamic inheritance laws and the third is prohibition of interest on capital. The class limitations of these conceptions of Islamic socialists are clearly reflected on their handling of the problem of private property. To conclude author says that for all disparity between national socialist theories and scientific socialism, the former are often of great positive value objectively in so far as they pave the way for the progressive development of the newly independent countries.

213. MINTJES (H). Debate on Islamic socialism in Pakistan. Part 1 and 2. *al-Mushir'* 20, 1-2; 1978; 38-72.

The main question treated is whether there is such a thing as Islamic socialism and if there is, how it differs from other forms of socialism. In part I the author deals with; 'Early Islamic socialism (to begin with al-Afghani, 1839-79) Indian Muslim authors with socialist sympathies between the two World Wars (Ubayd Allah Sindhi, Mohammad Iqbal) the view of some Pakistani thinkers after partition (Abdul Hakim, Ghulam Ahmad Parvez, Abul Ala Mawdudi), and the factual developments in Pakistan since 1947 (resulting in the 1979-victory of the Pakistan People's Party, which had made 'Islamic Socialism' its political slogan). The second part analyses the various concepts of socialism in the Islamic world; the main question is why Islamic socialism should be called socialism. At least in Pakistan most of the theorists of Islamic socialism wanted a home spun variety of socialism, based upon the *Quran* and the *Sunna*. The author further deals with the political developments in Pakistan in the late 1960s and their impact on the debate on Islamic socialism. The 1970 victory of the socialist parties did not mean that the electorate had opted for socialism: it did mean that they perceived what lived among the masses.

214. MINTJES (H). Debate on Islamic socialism in Pakistan. *al-Mushir*; 20, 4; 1978; 152-169.

(for the first two instalment see entry No. 213.)

Having described the historical background of the debate and some efforts made by Muslims to justify and realise an Islamic socialism the author draws here some conclusions in the light of the observations of certain Muslim authors. Among others, it is concluded that, after the Bhutto era and the emergence of the Nizam-i-Islam movement, to say that Islam is strongly socialist seems to have become almost sacrilegious. Because of its association with Bhuttoism 'Islamic Socialism' has become a bad word in Pakistan. The feeling of frustration and crises after the Bhutto era intensified the zeal for an Islamic order and speeded up the process of working seriously for its materialization. "Muslim have to give an Islamic answer to the problem of modernized poverty and not simply borrow from other. In fact, so we think this was exactly the underlying idea of the original theory of Islamic socialism." It will depend on the extent to which the Nizam-i-Islam will realise the cherished goal of "redressing the need of the toiling masses," whether the debate on Islamic socialism in Pakistan has stopped definitely or only temporarily.

ISLAMIC FUNDAMENTALISM, LAWS

215. CAMPBELL (Robert B). Restoration of Islamic law Vs. Secularization: Three law conferences in the Middle East-1976. *Cemam Report*; 4; 1976, Pub. 1978; 7-24.

Discusses three major conferences on law which took place in the Arab world, two in Riyadh, Saudi Arabia, the other in Tunis. The striking contrast between the two Saudi gatherings and the one in Tunis "represents a concrete instance in the continuing tension in the Arab Muslim world between two traditions of law, a tension which basically bears witness to a straggle between the religious institutions and what must be called—but with caution -the forces of Secularization." "Despite the Gulf that separates the two traditions, a continual mutual influence continues to be felt wherever the practical question of new legislation arises." This paper first describes the conference on Islamic Penal Law and the subsequent conference on Islamic jurisprudence (Riyadh). Secondly, an account of the 13th annual Arab Lawyers conference (Tunis) points out the contrast between it and the Riyadh conferences. The Riyadh conference are one more indication of steady pressure in today's Islamic world to restore the Islamic Sharia. The Tunis conference, however, revealed a totally different area of concern. Neither in the subject discussed nor in the final resolutions is Islam even mentioned. The motivations are quite different. The Arab Lawyers expressed their legal concerns in terms of the universal character of human rights, in the rule of modern positive law, and the principles of brotherhood, solidarity and unity of purpose that make the Arab one nation.

216. QADRI (Moinuddin). Islamic law in the modern world. *In*

Anwar Moazzam, *Ed. Islam and contemporary Muslim world.* New Delhi Light and Life, 1981, P. 133-141.

"It is a good augury that revolution has already begun in some of the Muslim states like Iran, Pakistan and Libya, to set up a purely Islamic government to implement Islamic laws in all spheres of life - political, economic and social. It is evident from these facts that struggle for revivalism of Islam has changed into a movement and spreading all over the Muslim world. The time and circumstances afford the best opportunity for the concretization of the idea of central authority, unity and the implementation of Islamic law. The time demands that the Muslim scholars should concentrate all their energies on the reconstruction of Islamic thought and rectification of Islamic law on modern scientific lines. Concreted efforts of the Muslims in the right direction would surely accelerate the process of Muslim renaissance and herald a new era in the Islamic world".

217. SIDDIQI. (Mohd. Suleman). Concept of Hudud and its significance. *In* Anwar Moazzam, *Ed. Islam and contemporary Muslim world.* New Delhi, Light of Life, 1981, P. 151-80.

Author says that practical implementation of *Shariah* in general and *Hudud* in particular has taken a greater significance in Iran, Saudi Arabia and Pakistan which making efforts to implement them. Most of the Western and also some Muslim scholars seem upset that the current religious fundamentalist revivalism will push the Islamic world back into the dark ages. According to Quran and the Sunna of the Prophet the crimes falling under hadd are five (1) adultery (Zina) (2) Theft (Sarqa) (3) Alcoholism (Shurb-e-Khamr) (4) False accusation of adultery (Qazaf) and (5) Highway Robbery (Qata-e-Tareeq). The moral, judicial and philosophical significance of had penalties are based on the concept of punishing those who have violated the rights, the dignity or the property of other individuals or society in general. They are meant to help evolve a just Islamic order. The rich and the poor, free Muslim and slave in the Muslim polity deserve and receive the same treatment alike. this in turn, suggest that the concept of hadd was tailor made to help bring about the kind of society and the moral order that the religion of Islam foresaw. The religion of Islam also lay great emphasis on mercy. Thus had reinforces the central thrust of Islam which is that the laws of the Quran and Shariah are the guiding spirit throughout the life of a Muslim.

218. WASSEL (Mohammad). Islamic law, its application as it was revealed in the Quran and its adoptability to cultural change. *Hamdard Islamicus;* 6, 1; 1983; 53-61.

The article includes and discusses, from an Islamic standpoint; (1) comparison between Judaism, Christianity and Islam; (2) Muslim stance towards the teachings of the Quran and the Sharia; (3) the situation in Muslim countries; (4) adaptability of the Shariah to the twentieth century; (5) the establishment of the Islamic state.

ISLAMIC FUNDAMENTALISM, LAWS, INDIA

219. PFEFFER (George). Muslim law and customary law of Muslims in South Asia: Two cases of adaptability and deviation. *Al-Mushir;* 22, 1; 1980; 4-16.

Author "identifies and sets in their respective contexts the different patterns of inheritance which obtain among Punjabi agricultural tribes and islanders from India. Both types of customary behaviour are contrary to Islamic law, the *Shariat.* This alerts us to an inevitable tension which exists between a religion's declared universality and its actualization in widely different cultures." In his conclusion, the author emphasises that "these problems may not be overlooked, least another, more disturbing problem, namely fiction, is created. These are bound to arise among Muslim because of the universal mission Islam claims to have, e.g. to an American docker, and Arab herdsman, an African planter, a Philippine hunter, or a German merchant. Each of these, after having accepted Islam, may commit the folly of denouncing the other on account of his different cultural background. Then, however, they would have to ask themselves; who is a Muslim for that is, after all, the real thing that matters. The answer would be unanimous. Any person acknowledging the unity of God and the Prophethood of Muhammad."

ISLAMIC FUNDAMENTALISM, LAWS *in relation to* SOCIETY

220. MUSLEHUDDIN (Muhammad). Islamic law and social change. Islamic Stud; 21, 1; 1982, Spring; 23-54.

Deals extensively with topics like: Law and conflicting interest, various kinds of pragmatic positivism, essence of human law, theories of social change, etc. In the second part, turning to Islamic law, he contrasts that law (unerring, divine in origin and, therefore, perfect and for all time) with human law, which is the product of reason, subject to change and liable to err. He also criticizes orientalists who, as it seems, "have launched a campaign to subvert Islam by their unfair criticism of Islamic law." The author finally considers the broad general principles of the Quran (which though immutable, are able to meet the growing needs of society and Islamic law) and the so-called rule of necessity and need. Islamic law is "capable of reconciling stability with change. It is the only law which effects a conjunction between these two conflicting tendencies and makes them work in unison. Thus the problem is solved."

ISLAMIC FUNDAMENTALISM, LAWS in relation to SOCIETY, EGYPT

221. MARTIN (M) and MASAD (RM). Restoration of Islamic law vs Secularization: Return to Islamic legislation in Egypt. *Cemam Reports*; 4; 1970; 47-78.

Already for over a century Egypt has been applying modern Western codes in most areas of law, and yet this experience has to prevented the rise of an urgent demand to revise the law by returning to the application of the Islamic Sharia. This article firstly gives a chronicle, in order to demarcate the sequent of recent events in the movement to return to the *Sharia*, it further describes the actors and the interpreters of the movement and analyses the discussion that has taken place within each of the groups taking part in the extensive national debate. The author thus determines those positions which are properly Muslim vis a vis the process of secularization which is affecting the entirety of modern religious society. Finally some remarks are devoted to the repercussions of secularization on the minds and thoughts of the people involved. "Whether exactly understood or not, the secularised society of the western type is rejected by all." "If then, sociologically speaking, a movement of secularisation takes place, in fact, in Muslim society, it is not recognised as such. All change is integrated or re-integrated with religion." "The Azharites jurists, and intellectuals are only reflecting various strands of public opinion which express different religious points of view springing from varied educational background." The article is followed by "the constitution for the return to Islamic legislation."

ISLAMIC FUNDAMENTALISM, LAWS, PAKISTAN

222. ISHAQUE (Khalid M). Islamization of laws in Pakistan: Problems and prospects. *Bull Christian Inst Islamic Stud;* 1, 3; 1978, Jul-Sept; 24-29.

Deals with the present legal structure of Pakistan, which consists of several kinds of laws, and the effort to Islamization of these present laws. He says that "one of the objectives of the Pakistan movement was that Islamic legal system should replace the existing system which had room for Islamic law. But for the replacement of the current legal structure with the laws of Sharia had many problems. Every regime in Pakistan promised for the Islamization of laws but the position remained the same. Even the last constitution gave the government a period of seven years for Islamization of laws. However matter are not at a stand still. Some reforms have taken place and some are in the process of being adopted. The legislature have also been moved by the Council of Islamic ideology to make certain changes of other field to bring the laws in consonance with the Quran and the Sunnah.

ISLAMIC FUNDAMENTALISM, LAWS, SHARIAT

223. KHURRRAM JAH MURAD. Sharia: The way to God *Islamic Order:* 5, 1: 1983: 45-48.

"In the whole movement of Islamic resurgence, nothing emerges as more symbolic of Muslim aspiration then the commitment to establish the *Sharia.*" "The *Shariah* literally means a clear path. It is the path that man, in Islam must walk as he toils and strives to reach his creator." The author says that "to understand the whole basis and concept of the *Sharia,* one must understand the relationship between man and God that Islam lays down." The above article is the expression of the same feeling.

224. NAQVI (S Ali Raza). Prophetic *Sunna* in the Islamic legal framework. *Islamic Stud*; 19, 2; 1980, Summer; 120-33.

Discusses the importance of *Sunna* and says "in modern age the Muslim community is confronted with new perplexing and intricate problems and questions. Concrete basis for their answer can be found in the Quran and the Prophet's Sunna." It would be necessary for all the followers of the various Muslim schools to join hands and "make serious endeavors for the compilation of the most authentic and reliable traditions of the Prophet from the whole corpus of Hadith available today." "This tremendous task can only be accomplished by the sincere and unbiased efforts of all the Muslim scholars belonging to the various schools of Muslim jurisprudence." In the end the author says "instead of being condemned as being engaged in mere mimicry of the past, let us appeal to the collective conscience of the community to find out suitable answer to the present problems and questions."

225. TANZIL-UR-REHMAND. Application of Shariah in the Muslim world. *Radiance*; 17, 33; 1981, Dec., 27; 5, 11.

Study of Islamic legislation makes it clear that (a) in all Muslim countries except Saudi Arabia and to some extent Indonesia, there has been codification and state legislation of Islamic law. (b) the courts have enforced the Islamic law, so codified and legislated (c) all the countries except Sudan, have appointed commission as a step towards state legislation of Islamic law (d) the commission so appointed did not strictly adhere to the dictates of only school of *fight*. (e) lastly, it is also evident from the above survey that matters falling within the domains of Muslim personal law occupy dominant position almost in every Muslim country and the larger issues in the areas of socio-economic and constitutional supremacy of *Shariah* does not seem to attract these countries with the result, that the fields of constitutional law, commercial law, industrial and labour law and international law are almost untapped and no serious efforts for their condification and inforcement appear to have been made.

ISLAMIC FUNDAMENTALISM, LAWS, SHARIAT, SUDAN

226. al-TURABI (Hasan Abdullah). On recent Shariah enact-

ments in Sudan. *Radiance*; 19, 29; 1983, Nov. 27-Dec, 3; 5.

The news of recent Shariah enactments in Sudan has raised a controversy within and outside the country. In Sudan while the Al Ansar led by the former Prime Minister, Sadiq al-Mehdi, has denounced it as exploitation of Islam, the *Ikhwanul Muslemeen*, cooperating with the Numeiri regime since July 1977 consider it a positive step towards Islamization. In this article the author has put forward his views regarding this issue i.e. the *Shariah* enactments in Sudan.

ISLAMIC FUNDAMENTALISM, LAWS, TRENDS

227. LIEBESNY (Herber J). Judicial systems in the near and Middle East: Evolutionary development and Islamic revival. *Mid East J*; 37, 2; 1983, Spring; 202-17.

"One of the present day phenomenon in the Islamic world is political pressure for a return to a strict application of Islamic principles in legislation and judicial practice. In Iran and Pakistan the dominant political leadership.....has espoused this trend and given it a constitutional foundation. In other countries such as Egypt, no basic changes have been made in the legal and judicial system to date. In a third set of countries, notably Saudi Arabia and Afghanistan before the Communist coup, a dichotomy developed, with Sharia courts applying religious laws and special commissions or courts applying statutory rules enacted to meet the demands of modern times." The article discusses these dominant trends in legal development, assessing the emergence of comprehensive codification, the assessing role of the Islamic jurist, the Mufti, and the impact, if any, of Islamic fundamentalism on the legal system.

ISLAMIC FUNDAMENTALISM, LEADERS, EGYPT, HASSAN AL-BANNA

228. HABID AHMAD (M). Imam Hassan Al-Banna and the call of the 'Muslim Brotherhood'. *Muslim Digest*; 28, 7; 1978; 2-15.

This is a translated chapter from the late author's book on "The Renaissance of the Muslim people". At the end of the thirties the call of the 'Muslim Brotherhood' movement merged in Egypt, a call which is Islamic in base, aiming at general revival, with a distinct upto date style and univer-

sal directions. The movement was founded (1928) by Imam
Hassan al-Banna, whose educational background, further
life and significance for the 'Ikhwan-al-Muslimoon' or the
Muslim brotherhood society is briefly sketched out. The
chiefs of the society devoted themselves to the fulfillment
of the objectives of the call claiming ideal programme for
prosperity and liberation. They established centres for
youths in an attempt for them to adhere to the Islamic traditions and cost off imported customs and habits which are
not compatible with the Islamic religion. In each of these
centres there existed the systems of families and spiritual
troops; also clubs and camps for youths. Hospital and
clinics were also established. A daily newspaper expressing
their reforming principles was issued. The Arab nations
were flooded with their publications, intended to reveal
their call in the religious, cultural and social fields. The
society was oppressed severely by the British and Egyptian
authorities. In 1949 al-Banna was killed in Cairo. The most
important objective of the society remains the establishment
of Islamic governments with the Quran as the governing
constitution.

ISLAMIC FUNDAMENTALISM, LEADERS, EGYPT, SAYYID QUTB

229. HADDAD (Yvonne Y). Sayyid Qutb: Ideologue of Islamic
revival. *In* John L Esposito, *Ed. Voices of resurgent Islam.*
New York, Oxford Univ. Pr. 1983: P. 67-98.

Perhaps the one of the most influential figures in contemporary Muslim revivalist thought is Sayyid Qutb of Ikhwan
al-Muslimoon (the Muslim Brotherhood). His books have
been translated and circulated widely throughout the Islamic world. His interpretation of Islam inform much of
contemporary Islamic revivalism. Author in the present article demonstrate why Qutb is a model for a process common to many Muslim revivalists. Born in a traditional
Egyptian village, his early traditional upbringing was followed by exposure to and enchantment with the west.
However he became disaffected with the west. He returned
to Islam convinced that only an Islamic alternative could
provide the ideology and values so sorely needed by Muslim society. Qutb joined the Muslim brotherhood and spent
the remainder of his life as an Islamic activist. Imprisoned
for ten years and finally executed by Nassar in 1966, Sayyid

Qutb has since that time been known as "the martyr" (Shahid) of the Islamic revival.

ISLAMIC FUNDAMENTALISM, IRAN, ALI KHAMENEI

230. HALLIDAY (Fred). Embodiment of fundamentalism. *Times of India*; 1981, Oct., 20; 8.

Ali Khamenei is a man who embodies the core of the Islamic revolutionary movement. It was people like khamenei who organised the mass protest that destroyed the Shah's regime in 1978. When the war with Iraq broke out in September 1980, Khamenei began visiting the front and, would urge the soldiers to fight against infidels and enemies of Islam.

He was wounded in a bomb explosion in June and has been relatively out of action since, but his emergence as president marks the final departure from Khomeini regime of those lay Muslim radicals were so prominent in the early days after the Shah's fall. This article is the expression of the above feeling.

ISLAMIC FUNDAMENTALISM, LEADERS, PAKISTAN, MAWDUDI

231. ADAMS (Charles J). Mawdudi and the Islamic state. *In* John L Esposito, *Ed. Voices of resurgent Islam*. New York, Oxford Univ. Pr. 1983, P. 99-133.

Author points out that like Sayyid Qutb, Maulana Abdul Ala Maududi may be viewed as an ideologue of a resurgent Islam. In 1941 he founded a Jamat-e-Islami a Muslim religio-political movement intimately involved in South Asia. Author in the present article shows how Maududi responding both to this South Asian experience as well as general conditions of Muslims throughout the world, called upon Muslims to restore Islam's primacy in their personal as well as their political lives. Convinced that Islam is a comprehensive way of life, he wrote extensively on nature and character of an Islamic state and through the Jamaat-e-Islami worked to implement politically the Islamic ideal.

232. BROHI (Allahbuksh K). Mawlana Abul Ala Mawdudi: The man, the scholar, the reformer. *In* Khurshid Ahmad and Zafar Ishaq Ansari, *Ed. Islamic perspectives*. Leicester, Islamic

Foundation, 1979, P. 289-312.

In the present article the author has tried to give an outline of the historical role that Mawlana Abul Ala Maududi has played in 20th century. He has attempted to highlight only that aspect of his (Mawdudi's) activity which has a bearing on what might be called his commitment to secure the moral, spiritual and intellectual regeneration of the people of Pakistan in particular, and the Muslims all over the world in general. "He has endeavoured to the best of his ability to secure the establishment of a model social order which may reflect the values and promote the ends for which, according to teaching of Islam, man has been created by his creator."

233. KHURSHID AHMAD and ANSARI (Zafar Ishaq). Mawlana Sayyid Abul Ala Mawdudi: An introduction to his vision of Islam and Islamic revival. *In* Khurshid Ahmad and Zafar Ishaq Ansari, *Ed. Islamic Perspectives*, Leicester; Islamic Foundation, 1979, P. 359-83.

The two ends of the twentieth century present two different pictures of the Muslim world. At the beginning of the century, the Muslim were in a state of disarray. Now in the last quarter of the twentieth century the Muslim world presents a some what encouraging picture. The chains of political slavery in many parts of the Muslim world have been shattered. The balance of economic power is an increasing desire to draw upon the intrinsic resources of Islam to build a new order. What lies, to a large measure, at the root of these political, economic, cultural and intellectual manifestations of resurgence is a rediscovery of the relevance of Islam to the problems and challenges of the time. This confidence and vitality are reflected in the movements of Islamic revival which have emerged in different parts of the world during the last fifty years. One of the chief architect of this movement is the quite and unassuming thinker, reformer and leader Maulana Sayyid Abul Ala Maududi. This paper attempts to present a systematic introduction of Maulana Mawdudi's thought, and the movement which has arisen on its basis. This has been prefaced by an extremely brief sketch of his life, mainly to serve as the background against which his ideas may be better appreciated.

234. MINTJES (H). Maulana Mawdudi's last years and the resurgence of fundamentalist Islam. *al-Mushir*; 22,2; 1980; 46-73.

The article concentrates on Mawdudi's last three years and on his influence during this period on developments in Pakistan and in the Muslim world at large. It first describes the coming to power of Mawdudi's organisation (Jamaat-i-Islami) in 1978, the specific Islami (Socio-economic) decisions linked up with that, and the direct bearing of Mawdudi's writings on present day affairs. Mawdudi was one of the greatest spokesman of Muslim fundamentalism in the 20th century, whose main contribution was his fresh interpretation of Quran which helped create a renewed confidence "that Islam is not something outdated and absolete but a living force relevant also to 20th century circumstances." His significance can also be seen from his international prestige. The article goes more in detail into Islamic fundamentalism, "which is represented best by Mawdudi's Jamaat in Pakistan and the Muslim Brotherhood in Egypt". and "is a movement some where in between westernizing modernism and *Ulamas* traditionalism." Mawdudi's prestige also becomes evident from his special relations with present-day's Iran. Fundamentalist thinking is "the real moving force behind the current wave of Islamization in the Muslim world", which is further illustrated here by description of recent development in Pakistan. The article finally points to Mawdudi's lasting significance, which "lies more in removing doubts and misgivings about Islam and stiffening Muslim self-confidence than in offering detailed suggestions for an Islamic order."

ISLAMIC FUNDAMENTALISM, LEADERS, SUDAN, MAHDI

235. VOLL (John). Sudanese Mahdi: Frontier Fundamentalist. *Int J Mid East Stud*; 10; 1979; 145-66.

The Sudanese Mahdi (Muhammad Ahmad al-Mahdi) has been pictured as a villain, as a hero, as a reactionary, as a revolutionary etc. The resurgence of Islamic activism in recent years gives added importance to efforts to understand the fundamentalist tradition in Islam. An analysis of the Sudanese Mahdiyya and its place in the tradition can provide insight into both the dynamics of the Mahdiyya itself and some aspects of Islamic fundamentalism. This requires that the Mahdiyya be examined in the broader context of the whole Islamic experience. Many writers use the title 'al-Mahdi' as the starting point for their analysis of Mohammad Ahmad's place in Islamic history. In this way they start with and describe the *Shia* concept of the Mahdi. This *Shia* context points the analysis a way from Islamic fundamentalism which usually was opposed to *Shia* tendencies. It is clear, however, that the distinction between the *Sunni* and *Shia* concepts of the Mahdi must be kept in mind in analysing the Sudanese Mahdiyya and that the concept utilized there was clearly within the *Sunni* and not the *Shia* tradition. This analysis involves three interlocking assumptions which are clerified in applying it to the Sudanese Mahdiyya. First, it assumes that there is such an analytical entity as the "Islamic fundamentalist tradition." Second, it asserts that at least one kind of Mahdism fits within the frame work of that tradition, and finally, that Muhammad Ahmad al-Mahdi in the Sudan was "that kind of Mahdi."

ISLAMIC FUNDAMENTALISM, LIBYA

236. ANDERSON (Lisa), Qaddafi's Islam. *In* John L Esposito, *Ed. Voices of resurgent Islam.* New York Oxford Univ Pr. 1983, P. 134-49.

From a Western perspective, the name most commonly associated with Islamic politics is probably Libya's Muammar Qaddafi. The author in the present article assesses the Islamic character of Qaddafi's Libya. Colonel Qaddafi's early appeals to Islam shortly after seizing power, the introduction of Islamic laws, his attempt to follow in the footstep of Nasser as an Arab leader, and his use of petrodollars to extend Libya's influence in the Muslim world would seem to support assumptions that his is an Islamic government. However, while the west continues to view Libya as an Islamic state, to what extent is that really true today? What is Qaddafi's Islamic legacy?

ISLAMIC FUDAMENTALISM, MAHDISM, AFRICA

237. CLARKE (P B). Islamic Millenarianism in West Africa: A revolutionary ideology. *Relig Stud*; 16, 3; 1980; 317-339

Social and political scientists, historians etc. have widely differing views concerning the character of Islamic millenarian and/or Mahdist movement In Africa. The author examines one aspect of Islamic Millenarianism in the West African context: its allegedly revolutionary character. He concludes that there is a little evidence to suggest that the Islamic Millenarian movement in West Africa sought revolutionary change in the Marxist sense. Often the Mahdist idea was used as an instrument against governments which were seen as oppressive and unjust. During the colonial era Mahdism become the tool of a counter-religious and cultural ideology. It was also used, exceptionally, as a device for integrating Muslim and no-Muslims. No where is evidence to suggest that Mahdism sought an radical, revolutionary transformation of society by means of the class struggle. In terms of economic change, Mahdism had little to say except that the directive of the Quran and the Sharia should be followed with regard to alms-giving, inheritance etc. In some respects the Islamic millenarian vision in the West African setting has more in common with the German idealism of Hegel and with Rousseau's ideas contained in the social contract than with Marxism as a theory of revolutionary change.

ISLAMIC FUNDAMENTALISM, MAHDISM, SUDAN

238. KAPTEIJNS (L E M). Religious background of the Mahdi and his movement. *African Perspectives*; 2; 1976; 61-79.

The emphasis in this article is on the religious motivation of the Sudanese Mahdi (1881), rather than on the social, economic and political causes of the Mahdiyya as a movement. The author thinks that these causes have been over emphasised at the expense of the religious factor, which is "not simply a way of expressing underlying, social and political convictions". She wants to demonstrate how the Mahdi fits into the Islam of the time, in which since the early 19th century a revival took place. The Mahdi is the divine leader chosen by God at the end of the world. His mission is to establish theocracy and so to re-enact the life of the original community of Islam. Each time narrowing her scope, the author subsequently discusses the context of world Islam, Muslim West Africa and (most extensively) Sudanese Islam and Sufism especially. The Mahdi did not have nationalist motives. He was committed to the true Islam of the Prophet's time.

239. MOHAMMED SADEQ. Sudanese Evolution: From crucible to centrifuge. *Radiance*; 15; 1971, Oct, 24; 5, 12.

Author in this article says that the Mahdists aimed at bringing back the rule of the Koran and the Sunnah. The Mahdi of Sudan was influenced by the teachings of Ibn Taymiyya and held views similar to those of Mohammed Ibn Abdul Wahab in Saudi Arabia. He carried out many reforms in all aspects of the life of the people. For the first time Muslim tribes acted together and worked together in available political unity, transcending traditional bonds. Many people gave their life for the cause, and all the Sudan came under Mahdi's rule. The British delt severely with all attempts made for any religious revival.

240. MURAD (Hasan Qasim). Mahdist movement in the Sudan. *Islamic Stud*; 17, 3: 1978; 155-84.

The re-discovery of manuscript material on the Mahdist movement in the Sudan (Mahdiya), though not yet fully exploited as well as to the re-awakening of the interest in the movements resulting in brilliant, careful and judicious scholarship by persons like Hold, Shibeika, Theobald and Hill, students are now in a better position to understand *Mahdiya*, not simply as an episode or interlude in Egyptian or British imperial history, but as a movement in its own right as an "autonomous historical process" seen against its Sudanese African as well as its wider Islamic background. The author quotes P M Holt (Mahdist State) who pointed out that the Mahdiya was a movement of religious origin which was assisted in its development by political social and economic stresses in Sudanese society and which accomplished a political revolution - the over throw of Egyptian rule and the establishment of an indigenous Islamic state". This article deals with the first phase of the Mahdiya, that is Mahdya in its formative stage, as a movement of religious protest against "internal decline and external encroachment" symbolized mainly in the (Un-Islamic) character of the Khedival administration in the Sudan.

ISLAMIC FUNDAMENTALISM, MALAYSIA

241. von der MEHDEN (Fred R). Islamic resurgence in Malaysia. *In.* John L Esposito, *Ed. Islam and development.* New York, Syracuse Univ. Pr. 1980, P. 163-80.

Examines the renewed emphasis on the identity of Malay ethnicity and 'Islam and its impact on Malaysian society. The author dates the beginning of the resurgence of Islam in Malaysia to the governments response to communal riots in between Malays and Chinese. The process of Islamic revival has been further enhanced by closer ties with the Middle East and the growth and impact of Islamic missionary movements. The later are especially important since their goals are not only the conversion of non-Muslim but also religious renewal among Malay Muslims. Author says "some observers believe that some eighty percent of Malay university students are part of one of the many Dakwah organisations." This phenomenon demonstrates the ability of modernization to contribute to Islamic renewal rather than to weaken religion through secularization. The author assesses the socio-political implications of the Islamic resurgence in Malaysia.

242. NAGATA (Judith). Islamic revival and the problem of legitimacy among rural religious elites in Malaysia. *Man*; 17, 1; 1982; 42-57.

The religious Ulama of Malay Islam have a long history of attachment to village social life. Their roles as teachers and a respected scholars, reinforced by strategic marriage alliances and aristocratic patronage, made them a local rural elite, although in matters of orthodoxy they have absorbed a number of non-Islamic village customs. The recent urban and middle-class Islamic revival (Dakwah), centred largely on a youthful, university and foreign educated intelligentsia, has opened a challenge to the authority of the *Ulama*, although the former lack of traditional religious credentials, combined with high secular status, creates some ambivalance. With Islamic revival missionaries now preaching in the countryside, the *Ulama* are trying to steer a middle course between accommodation to new ideas and preservation of their legitimacy.

ISLAMIC FUNDAMENTALISM, ORGANIZATIONS, IKHWAN AL-MUSLEMOON, EGYPT

243. BELLO (Iysa Ade). Society of the Muslim brethren : An ideological study. *Islamic Stud*; 20, 2; 1981; 111-27.

The Society of Muslim Brethren, founded in Egypt in 1928

by Hasan al-Banna, still plays a conspicuous role in contemporary Islamic history. This article summarizes some aspect of the society's ideology. It draws "its sources mainly from the writings of the founder of the society and from the writings of some of its leading spokesmen for the sake of clarity whenever conditions demand it." It is concluded "that ideology of the society of the Muslim Brethren can be summed up in this short sentence. Islam is a way of life. The originality of the Brethren lies not in their ideology, many elements of which may be found in the preaching of Jamal al-Din al-Afghani, Muhammad Abdul, Rashid Rida, Muhammad Ibn Abd-al-Wahab, and others. It lies in the fact that their founder, by shifting the teachings of these predecessors, made a critical selection from them, simplified it, made it stricter, and above all made it the living ideology of a popular powerful movement. His teachings, as he mentioned, were not a innovation but a renovation. The idea of the Muslim Brethren were and are still widely spread in the Muslim world. They have a remarkable influence on religious writings, which they enriched with their prolific writings."

ISLAMIC FUNDAMENTALISM ORGANIZATIONS, IKHWAN AL-MUSLEMOON, SYRIA

244. ASIF (A U). Syria and the Ikhwans. *Radiance*; 17, 44; 1982 March, 14; 12.

The Ikhwanul Muslemoon in Syria are suffering a great deal at the hands of the oppressive regime of President Hafes al-Asad. They are being subjected to great atrocities by Bathist Nusairi rulers of Syria who although in a minority hold the reigns of power and are averse to the efforts at Islamic revivalism spearheaded by the Ikhwans and other Islamic groups.

ISLAMIC FUNDAMENTALIS, ORGANIZATIONS, JAMAAT-E-ISLAMIC, INDIA

245. ABDUL MOGHNI. Resurgence of Islam and Jamaat-e-Islami. *Radiance*; 16, 40-41; 1981, Feb, 15-22; 15.

Jamaat-e-Islami was founded in the year 1941 by Allama Abul Ala Mawdudi. His voice had been calling people towards the straight path of life through his scientific and scholarly thesis and treatises. His writings had created a circle of devoted workers in the way of Islam. Now the world wanted a powerful organized action to combat the powerful forces of disintegration and destruction. Hence the establishment of Jamaat e-Islami. The party of Islam-as the centre and instrument of an Islamic movement. This movement sought to project Islam as a universal ideology, meant for the whole human race rather than a religion. Maulana Abul Ala Mawdudi aimed at a radical and total change in the whole prevailing set-up of the modern society as well as state, on the basis of the fundamentals of Islam. In the thirties, he prepared the ground with his writings, for a full fledge Islamic movement. Now the Jamaat-e-Islami has become more of India is creating pockets of influence among the non-Muslim as well. In Pakistan the Jamaat-e-Islami has not only polarised the people of that country for Islam, but made an impact also on the whole of West Asia. The struggle for the inforcement of Islamic laws in Pakistan, waged by the Jamaat, has set a model for the Islamic movements in Muslim countries. It now well be said that resurgence of Islam in modern world has been possible largely due to the efforts of Jamaat-e-Islami.

ISLAMIC FUNDAMENTALISM, ORGANIZATIONS, REPUBLICAN BROTHERS, SUDAN

246. MAGNARELLA (Paul J). Republic Brothers: A reformist movement in the Sudan. *Muslim Wld*; 72, 1; 1982; 14-24.

Author discusses "The Republican Brothers'. a Sudanese reformist movements. He "points out to the modest rise of the movement's membership. Moreover, it is obvious that many of their teachings are unacceptable to most other Muslims." The most radical doctrine of the movement (also called the New Islamic Mission) is that the Quranic revelation has two devine messages; the first and the second, based respectively on the Medinese and Meccan texts. "One can detect many similarities between the ideology of the Republican Brothers and those of others Muslim reformers: One being the rejection of traditional interpretation of the Quran by the Ulama, an other being the disbelief in hell of in an eternal damnation in Islam. The New Islamic movement deems the entrenched Muslim authorities outmoded and irrelevant to the present age. The second message of Islam is a frontal attack on the establishment."

ISLAMIC FUNDAMENTALISM, ORGANIZATIONS, TABLIGH

247. JANSEN (Godfrey). Towards an Islamic society. *Seminar*; 290; 1983, Oct; 33-36.

According to the author "the largest and least known religious movement in the Muslim world to day is the Tabligh (Revival) movement which, though it is itself non-militant, provides the militants in Pakistan and Bangladesh with a solid foundation of revivalist Islam. It was started in Delhi around 1945 and developed in the Islamic equivalent of moral rearmament. Its aim is to produce good Muslims, born-again Muslims. The Tabligh movement has no regular office, nor oganizational structure, not even a fixed membership, yet it has attracted to its rank every sort of Muslim, from university professor to peasants. Tabligh, has spread its activities to every country in which Muslims live. The objectives is not an Islamic State but an Islamic society to be achieved through a slow transformation, through education and missionary endeavor, to convert Muslims in a society, that society will inevitably become an Islamic society. If that Islamic society establishes it-self and endures then, no less inevitably, it will produce an Islamic State.

ISLAMIC FUNDAMENTALISM, ORGANIZATIONS, WAHHABI

248. ZAHARADDIN (M S). Wahhabism and its influence outside Arabia. *Islamic Quart*; 23, 3; 1979; 146-57.

In this paper an attempt is made to study Islamic reform movements that was launched in the Arabian Peninsula in the eighteenth century. The movement began as a revolt against the then existing religious practices. This movement, or at least its influence, was not confined to Arabia alone, but reached as far west as North Africa in the first years of 19th century and as far South as the shores of India. The founder of this movement was an Arab called Muhammad Ibn Abd al-Wahhab. His movement is called *al-haraka al-Wahhabiyya* (Wahhabi movement or Wahhabism). "In the Islamic world the term Wahhabi has always been applied loosely to designate "any person who aims at religious reformation, even if that person be a complete stranger to Najd." Because in the modern period many reform movements have had a tendency to borrow from Wahhabis or use the same source material (i.e. the Quran and the Sunna)" The author says that "it should however be borne in mind that Wahhabism remains a significant current in contemporary Muslim thought."

ISLAMIC FUNDAMENTALISM, ORGANIZATIONS, WAHHABI compared with SANUSI

249. ANSARI (Zafar Ishaq). Some reflections on Wahhabiyah and Sanusiyah. *Islamic Order*; 4, 4; 1982; 58-65.

Points out that the phenomena of Islamic resurgence is not taking place in a vacuum. It has its antecedents. The movements of today are connected with similar movements of the past. The 20th century Islamic resurgence is related to those movements of Islamic reform and revival which stirred the minds and inspired the hearts of Muslim in different parts of the world in the 18 the and 19 the centuries. The author has discussed about the Wahabi and Sanusi movements in this article. He has also make a brief peripheral reference to another great figure of the eighteenth century-Shah Wali-Allah of Delhi.

ISLAMIC FUNDAMENTALISM, PAKISTAN

250. ANAND (Som). Islamism and Pakistan's social realities. *Man Develop*; 2, 1; 1980, March; 93-107

"Pakistan has become a focal point in the present wave of religious fundamentalism sweeping many parts of the Muslim world". Pakistan can last only as long as ideological foundations remain intact. The movement for an Islamic order reached its climex during the agitation by the Pakistan National Alliance (PNA) in 1977, but it was basically a political movement. While tracing the history of Muslim religious revivalisam in the Sub-Continent, the Jamaat-e-Islami is the most authentic voice of Islamic fundamentalism in Pakistan. Bhutto was in fact a radical Islamist. Though politically he (Bhutto) made every effort to curry favour with Saudi Arabia but his Islam differed widely from Saudi fundmentalism. Bhutto was one of the Muslim to favour some kind of socialist system as the basis of the country's economy. But "while analysing the hurdles which the General Zia faces one has to see the social realities in Pakistan. The country's social, economic and political life has all the complexities of a land which has been under the impact of industrialization and notions of a democratic polity of the western style. In the end author says that " a sucessful introduction of the Islamic laws will depend on the Ulama's capacity to relate Islam to the requirements of the present days Muslims."

ISLAMIC FUNDAMENTALISM, PALESTINE

251. ZAFARUL ISLAM KHAN. Resurging Islamic movements in Palestine. *Radiance*; 16, 50; 1981, April, 26; 12

There are clear signs of an Islamic awakening in the Muslim lands occupied by the Zionists - with the Islamic revival in the occupied lands, the number of Muslims attending mosques for *Salah*, growing beards and the number of students in Islamic religious schools has increased to an extent that has alarmed the occupation forces which say many secret Islamic organization like Ikhwan, are operating in the occupied lands. "Religious awakening in the occupied territory should be viewed as a part of the developments taking place in the Islamic world".

ISLAMIC FUNDAMENTALISM, SAUDI ARABIA

252. OCHSENWALD (William). Soudi Arabia and the Islamic revival. *Int J Mid East Stud*; 13, 3; 1982, Aug.; 271-86.

Discussed about the phenomenon of religious revival and says that "it is easy to suggest the prospect of an Islamic revival in the Muslim nation states, but it is extremely difficult to specify the exact nature of such a revival, particularly in a country like Saudi Arabia. If Islamic revival means an increase in the power, prosperity and effectuality of Muslim States, then certainly Saudi Arabia in the 1970s has undergone an Islamic revival." "On the other hand, if Islamic revival means an intensification of the role of Islam in public life and in the hearts of men then little happened in this area, because Saudi Arabia's public and private life were already suffused with Islam."

ISLAMIC FUNDAMENTALISM, SENEGAL, HISTORY

253. MONTEIL (Vincent). V. Lat-Dyor, Damel of Kayor (1842-86) and the Islamization of the Wolof of Senegal. *In*. I M Lewis, *Ed. Islam in tropical Africa*. London, International

African Institute, 1980, P. 166-72.

The History of Lat-Dyor, Damel (King) of Kayor (1842-85), who was converted to Islam in 1864, demonstrates the factors which have determined the introduction and acceptance of Islam in Senegal, as well as the effects of Islamization on economic and political values and on the social structure. It throws light on the problem of contacts between Muslims and non-Muslims, it illustrates the confrontation at the end of the 19th century, it shows the dialectical relationship between Islam and nationalism.

ISLAMIC FUNDAMENTALISM, SOUTH EAST ASIA

254. WALKER (Dennis). Muslim movements in South East Asia. *Radiance*; 9, 46; 1972, May, 28; 7, 9

The organic bond which Malay Muslim nationalists see between their peoples and the larger world of Islam centered in the Arab Middle East implies that out side Muslim government could or should, become involved in Muslim insurgencies in South East Asia. If only in desperate quest for some international counterweight to the overwhelming means of repression which the non-Muslim governments under whom they live can command, Muslim dissidents in Thailand and the Philippines have worked to transform this possibility of international Islamic aid into reality.

ISLAMIC FUNDAMENTALISM, SPAIN

255. ABDUL HASIB. Spaniards re-discover the relevance of Islam. *Arabia*; 26; 1983, Oct; 70-71.

There are outstanding examples of the recent spread of Islam. The million of black Muslims in the U.S, the growing number of new Muslims in the United Kingdom, France and Germany-but the case of Spain arouses special interest. Spain was a Muslim country for more than eight centuries, its major language was Arabic, and its civilization produced men whose knowledge has been a major contribution to mankind. The author examines the Islamic revival in Spain-and the attention, not altogether sympathetic, it has attracted.

ISLAMIC FUNDAMENTALISM, TERMINOLOGY, CRITIQUE

256. NASR (Seyyed Hossein). Decadence, deviation and renaissance in the context of contemporary Islam. *In*. Khurshid Ahmad and Zafar Ishaq Ansari, *Ed*. *Islamic Perspectives*. Leicester, Islamic Foundation, 1979, P. 301-15.

Deplores the tendency towards ambiguity and careless use of these (decadence, deviation, renaissance and Islamic Fundamentalism) important terms by a large number of modernised Muslims and others. He refuses to believe that decadence, as a gradual process of ageing and of becoming more removed in time from the celestial origin of the revelation, took place in the Islamic world in the 7th/13th century as many western scholar maintain because cultural activity in some part of the Muslim world such as Iran and Pakistan (then India) did continue upto 12th/18th century. He stoutly defends the integrity of the Hadith literature and says that with the *Hadith* and *Sunna* even the message of the Quran would become incomprehensive to men. Islamic renaissance and Islamic fundamentalism to him means independence from the influence of the west and all that characterises modernism.

ISLAMIC FUNDAMENTALISM, THAILAND

257. ENGINEER (Asghar Ali). Islam in Thailan resurgence or consolidation. *Islam Mod Age*; 14, 1; 1983, Feb; 59-67.

Buddism continues to be the state religion in Thailand. Muslim constitute 4 % (1,85 million) of the population according to state statistics, Muslim sources claim higher figure). In any case, Muslims constitute the second largest religious group. This article gives some characteristic features of the Muslim community, which is well integrated with the Thai society. It concludes that the Thai Muslims, despite their cultural assimilation, are very proud of their being Muslims. Intellectuals increasingly become aware of their Muslim identity; some of them have come under the influence of political Islam of Khomeini. It is difficult to say whether they are fully aware of the implications Khomeini's Islam. The author's discussion with them "shows that they are more concerned with its symbolic and ritual aspects rather than understanding its deeper implications. The *Ulama* and the Muslim masses on the other hand hardly feel concerned with what is happening in Iran. They are more concerned with traditions of Thai Islam and summoning people to the mosque for ritual prayers."

ISLAMIC FUNDAMENTALISM, TRENDS

258. AGWANI (M S). Varied politics of resurgent Islam.
India Int Centre Quart ; 7, 3 ; 1980 , Sept ; 141-48

Discussed the phenomenon of Islamic resurgence in this article. When Shah of Iran left for Tehran, CIA had been ordered to conduct a world wide inquiry into the phenomenon of Islamic fundamentalism. As usual newspapers and periodical in the west picked up the theme and started publishing articles on "Islamic fundamentalism". Author questioned " why should development in one country (Iran) give rise to a sweeping generalization about the vast Muslim world? and tried to show the flaws in this particular thesis by giving examples of the diverse manifestations of Islamic revivalism; each manifestation often at loggerheads with the other. Author looking at forms of Islamic revivalism that not only serve the social and economic status quo but also pose a challenge to it. "Finally, the author believe that "it is quite natural for Muslims societies to invoke the norms of Islam, particularly in moments of crisis".

259. FARUKI (Kemal A)—Islamic resurgence; Prospects and implications. *In.* John L Esposito , *Ed. Voices of resurgent Islam.* New York, Oxford Univ Pr., 1983 , P. 277-291.

Emphasised that the religion has been part and parcel of Muslim history and politics, a religio-political movement with a universal mission to spread God's rule and his related way of life, the period of Muhammad, the early Caliphs, successive Muslim empires, and asorted movement to renew (Tajdid) and reform (Islah) the Islamic community are part of a continuance which extents to current and future movements in Islam. The author in the present article views current attempts at Islamisation both in relation to recent Islamic movement in Libya and Saudi Arabia and contemporary experiments in Pakistan and Iran. Such an evaluation is important because of the issues and questions that arise from such experiments as well as their potential implications for the future direction (s) of Islamic revivalism.

260. KHALID (Detlev. H) Phenomenon of re-Islamisation. *Mainstream* ; 17, 42 ; 1979 June, 16; 19-28.

In this article the author presents a perceptive analysis of the different trends in Islam the world over, including the continuous struggle between mystecism and formalism, the emergence of a 'third force' called fundamentalism or integratist Islam, the re-Islamization process observed in many Muslim countries, and the reaction of the forces opposing fundamentalism. The article raises and deals with aspects of continuing interest and importance and is highly thought provoking.

261. YAPP (M E). Contemporary Islamic revivalism. *Asian Aff*; 11, 2 ; 1980, June ; 178-95.

Author in the present article mentions some example to illustrate the extent and variety of the phenomenon of Islamic revivalism. In the field of social life we have seen that re-introduction of elements of Islamic law and the amendament of curricula in schools to give more time to religious teaching. In the politics Muslim political parties have achieved greater influence and secular parties have incoporated into their programme a variety of Islamic elements. There has been a notable increase in the activities and influence of secret or illegal Muslim organization. Author says that 'it appeared that the phenomenon of Muslim resurgence was not a matter of a few exceptional episodes which could be explained by special circumstances but that it demostrated the rejection of the secularist path by the greater part of the Muslim world; To find a satisfactory explanation of the phenomenon of Muslim revivalism was therefore, not merely a matter of practical importance, in order that a suitable tactical response by non-Muslims might be devised, but it was also a challenge to the Western concept of civilization and its development," says author.

ISLAMIC FUNDAMENTALISM, TRENDS, INDIA

262. ANSARI (Javed). Themes in Islamic revivalism. *Arabia*; 24; 1983, Aug; 32-33.

Author points out that in India Jamaat-e-Islami has mobilized popular support for its programme among both Muslims and non-Muslims. Among its critiques Maulana Zakaria of Jamaat-i-Tabligh has opposed Maulana Maududi in his book *Filna-i-Maududiat*. In this book Maulana Zakaria emphasised more on Ibadah while Maududi over emphasised the political aspect of the Islamic system. Critics of Maududi say his emphasis on political struggle means rejecting traditional values, but Javed Ansari argues that this is not the case. At present Tabligh Jamaat is facing stagnation., There is a need to revitalise the spiritual life of the Muslim masses and to make this revitalisation the basis of revolutionary Islamic political struggle. Some modern reformist movement assume that ideological accommodation is Permissible within all political systems. This makes it impossible to conceive of political initiative that can transform the nature of the Indian polity and makes the Ulema the natural ally of nationalists, Baathists and socialist regimes.

263. FAIZEE (Shameem). Changing tactics of Muslim revivalists. *Link* (Republic Day Number); 24, 24; 1982, Jan, 26; 33-34.

The Islamic Revolution in Iran and cry for Nizam-e-Mustafa in Pakistan started a wave of revivalism among the Muslims which was further strengthened with the massacre of Muslims in Moradabad in August 1980." "The mass conversion of harijans to Islam in Meenakshipuram has been yet another booster for the trend." "All this should be seen in the context of new tactice of various Muslim organization the formation of a new Muslim youth organization and writings in Urdu Press owned and read by Muslims." "In totality, the revival tendencies are being encouraged and we find the resemblance of the *Tabligh* and *Shudhi* movement of the late twenties in the wake of Meenakshipuram which naturally is an alarming sign."

264. MATHUR (Girish). Islamic fundamentalism and India. *Mainstream*; 19, 7; 1980, Oct. 18; 13-16.

The concept of Islamic fundamentalism have been borrowed by the Indian editors and commentators from imperial brains trust of the contemporary world, the U.S. Council on foreign relatins which employs it to find scapegoats for debacle of American foreign policy in west

Asia, South West Asia and the countries of South Asia with a high percentage of Muslim population. They ascribe all their misfortune to Islamic fundamentalism. They also see to justify the old policy by raising the bogey of Islamic fundamentalism which they say is a reaction to modernization encouraged by them in the affected countries. But to be a reaction to some thing which sought to be passed off as modernisation, Islamic fundamentalism has to be, by definition a historically modern phenomenon-which it is not. Historically, the rules of Islamic fundamentalism can be traced to the earlier days of Islam. The fundamentalist trend came to India alongwith the Muslim rulers and theologians attached to their courts. The fundamentalism in modern times, that is, in the age of imperialism, is a response to both the internal decay of Muslim societies and the external threat of imperialism. Those who have borrowed this concept of Islamic fundamentalism from its American user, will have to decide which fundamentalism they are talking about. Wahabi? The one that is developing in opposition to Wahabism? Ziaul Haq's? Ikhwan's? or one of those earlier ones like the *Ahle Hadith*? or the still earlier al-Ghazali's? All these fundamentalism cannot be put in the same basket. Neither any of the varieties of Islamic fundamentalism nor the idea of fundamentalism can serve as a tool of analysis for understanding what is happening in India. In any case the idea of Islamic fundamentalism being "abstracts" and intellectualised and devisive by nature, have limited appeal and cannot account for what has happened earlier in Kashmir valley or in other parts of Northern India recently.

ISLAMI FUNDAMENTALISM, TURKEY

265. ASIF (A U). On road to Islam from Kemalism. *Radians*; 17, 21-22; 1981, Oct, 4-11; 7-8.

This paper says that the current year has witnessed intensified struggle by the Islamic movement the world over. Turkey too is now under the sway of the tides of the Islamic resurgence movement. The Islamic movement which have received the mass support all over the country is actually the story of journey from Kemalism to Islam. The author has given the account as to how the Islamic movement in the country where every corner is full of status and pictures of Kamal Ataturk started. The author says that the military rulers realise that the Islamic revivalism movement can not be crushed as it has caught the imagination of the masses. The author also suggested that "It is time that West-oriented Kamalism is given a ceremonial burial which has done no good to Turkey and the Muslim Turkish people."

ISLAMIC FUNDAMENTALISM, WOMEN

266. HJARPE (Jan). Attitude of Islamic fundamentalism towards the question of women in Islam. *In* Bo Utas, *Ed. Women in Islamic Societies.* London, Curzon Pr. 1983, P. 12-25.

Author feels that instead to examine the points of view which exists and are propagated from within Islam, that is to say, how Muslim justify the women's position, rights and obligations with direct reference to Islam's norms and authorities. It is not sufficient to consider the factual status of women in one or another area. The intension here is rather, to take notice of how questions relating to women are understood, on the basis of the Quran and Muhammad's Sunnah, among contemporary fundamentalist.

267. SAMAR FATIMA. Nature and effects of the Islamic attitude to women, *Islamic Slud*; 21, 1; 1982, Spring; 105-121

Discusses several aspects of the change in the position of women, brought about by the reform of Islam. She concludes that Islam through the social system founded by Muhammad, "restricted unlimited plurality of wives, forbade female infanticide and discouraged divorce and limited the power of divorce possessed by the husband, guarded the rights of widows and female orphans, secured to wives of dower and enjoined kind treatment towards women whether mothers, sisters, daughters or wives, establish a law of inheritance and promised religious favour for both sexes on same levels."

ISLAMIC FUNDAMENTALISM, WOMEN, EGYPT

268. WILLIAMS (John Alden). Veiling in Egypt as a political and social phenomenon. *In* John L Esposito, *Ed. Islam and development.* New York, Syracuse Univ Pr. 1980, P. 71-85.

This article focusses on the motives and issues underlying an apparently in explicable regression. Egypt had led the

Islamic world in improving the status of women. Yet today middle class Egyptian women are donning Islamic dress. Is this simply a rejection of modernization in favour of Islamic fundamentalism? The author rejects such a conclusion. Rather, he sees its origin in a return to Islamic values. Faced with acute socio-economic problems, some women turned to their Islamic tradition for a greater sense of identity and authenticity. It is an attempt to develop a more indigenous, Islamic response in the midst of social disruption and a sense of spiritual malaise.

ISLAMIC FUNDAMENTALISM, WOMEN, LAWS

269. FAZLUR RAHMAN. Survey of modernization of Muslim family law. *Int J Mid East Stud*; 11, 4; 1980; 451-65.

Author starts by stating that "some strides have been taken to improve the status of women in Islam, but the weight of conservatism is still very strong." An important factor is that "Muslim societies still set a high value upon the family institution and assign a first priority to parental rearing children." In the author's opinion, "this factor is not likely to undergo much change-in fact, recently, a vehment vindication of this principle has been demonstrated in several Muslim countries with powerful stirrings in all of them— since it concerns an area of basic value of life." Since the forties "a rapid increase has occured in the education of women all over the Muslim world". The article considers the reformist enactments as they affect marriage, divorce, and inheritance and in the end says something about the relevance to the family institution of the family planning programmes officially undertaken in various Muslim countries.

ISLAMIC FUNDAMENTALISM, WOMAN, PAKISTAN

270. CARROLL (Lucy). Nizam-i-Islam: Processes and conflicts in Pakistan's programme of Islamisation, with special reference to the position of woman. *J Commonwealth Comp Polit*; 20, 1; 1982, March; 57-95.

Describes that General Zia-ul-Haq announced a programme designed to mould Pakistan into an Islamic Republic in fact as well as name. Nizam-e-Islam (the order of system of Islam) portends wide-ranging repercussions on economic and political life and institutions. The reforms are being introduced gradually. The purpose of this eassay is to hightlight some of the processes and conflicts involved in what is viewed by its designers as an Islamic evolution. In this article, author has focussed on family law and the position of women. Particularly against the background of events in neighbouring Iran, the rhetoric of Islamization has raised many questions concerning the place of women in the new order which author has tried to answer.

AUTHOR INDEX

TITLE INDEX

A

B

C

D

150